Why Society is a Complex Matter

Philip Ball

Why Society is a Complex Matter

Meeting Twenty-first Century Challenges
with a New Kind of Science

With a contribution by Dirk Helbing

Cover illustration: Credit: Franz Pfluegl/Shutterstock

ISBN 978-3-642-28999-6   e-ISBN 978-3-642-29000-8
DOI 10.1007/978-3-642-29000-8
Springer Heidelberg Dordrecht London New York

Library of Congress Control Number: 2012935045

© Springer-Verlag Berlin Heidelberg 2012

This work is subject to copyright. All rights are reserved, whether the whole or part of the material is concerned, specifically the rights of translation, reprinting, reuse of illustrations, recitation, broadcasting, reproduction on microfilm or in any other way, and storage in data banks. Duplication of this publication or parts thereof is permitted only under the provisions of the German Copyright Law of September 9, 1965, in its current version, and permission for use must always be obtained from Springer. Violations are liable to prosecution under the German Copyright Law.
The use of general descriptive names, registered names, trademarks, etc. in this publication does not imply, even in the absence of a specific statement, that such names are exempt from the relevant protective laws and regulations and therefore free for general use.
Product liability: The publishers cannot guarantee the accuracy of any information about dosage and application contained in this book. In every individual case the user must check such information by consulting the relevant literature.

Printed on acid-free paper

Springer is part of Springer Science+Business Media (www.springer.com)

Contents

Introduction VII
Society: a Complex Problem • Is Society
Predictable? • What is Complexity? • Modeling
Complex Systems • What this Book is for •
Further Reading

1 **On the Road: Predicting Traffic** 1
Flow States • All Together Now • Traffic Lights
Can Organize Themselves • Theory Into Practice
• Further Reading

2 **Every Move You Make: Patterns of Crowd Movement** 7
Walking on Computers • Panic • Better Rules,
Better Data • Planning Public Spaces • Further Reading

3 **Making Your Mind Up: Norms and Decisions** 13
All in This Together? • Peer Pressure •
Reputation and Trust • Further Reading

4 **Broken Windows: The Spread and Control of Crime** 18
On the Streets • Bad Influence • Hotspots •
Further Reading

5 **The Social Web: Networks and Their Failures** 23
Small Worlds • Clubs and Communities • When
Systems Fail • Further Reading

6 **Spreading It Around: Mobility, Disease
and Epidemics** 28
Getting About • Going Viral • Contagious
Behaviour • Further Reading

7 **After the Crash: Economic and Financial Systems** .. 33
The Problem with Economics • High
Expectations • The Bigger Picture • Further Reading

8 **Love Thy Neighbour: How to Foster Cooperation** ... 38
Resolving the Dilemma • Patterns of Niceness •
Reputation Matters • Crime and Punishment •
From Games to Reality • Further Reading

9 **Living Cities: Urban Development as a Complex
System** 43
City as Organism • Universal Maps • Planning
or Managing? • Further Reading

10 **The Transformation of War: Modelling
Modern Conflict** 48
The Power of War • Computerized Conflict •
Why Fight? • Further Reading

Summary 53

New Ways to Promote Sustainability
 and Social Well-Being in a Complex, Strongly
 Interdependent World: The FuturICT Approach 55
 Dirk Helbing
Why FuturICT is Needed • Why Information
and Communication Technology (ICT) is Crucial
• The Components of FuturICT • Towards
More Resilient and Sustainable Systems: How
It All Comes Together • Determining the
'Social Footprint' to Protect the Fabric of Society
• In Conclusion • Further Reading

Introduction

Society: a Complex Problem

It is becoming ever more clear that the twenty-first century is not a continuation of the twentieth, but something new. War is qualitatively different now from what it was half a century ago, and so is peace. So are consumerism, access to information, environmental change, health care, demography, and perhaps the very concept of democracy. It seems we are living not at the "end of history" after all but at the beginning of a new historical phase – one that demands new ways of thinking.

This is why it is time to escape the constraints of disciplinary thinking. The major challenges of the twenty-first century are not ones that can be understood, let alone solved, from a particular academic perspective. For example, if today's patterns of consumption make global mean temperatures destined to rise by even 2 °C, the consequences for international relations, biodiversity, food and water security, and human migration are immense, and yet are at this stage little more than informed guesswork. Simply comprehending and forecasting such impending crises, let alone mitigating them, is not just a question of having more accurate models of global climate, but must involve the integration of a host of socioeconomic, technological and political factors.

The most important novelty in the changes that are currently being felt by our societies and our environment stems from the profound impact of globalization: the linkages and interconnections that transcend states and societies. The interdependence of economies, cultures and institutions has become deep and dense, in large part thanks to the pervasive nature of information and communication technologies (ICT). Nothing will work that fails to take this into account: not the economy, not policing, not international diplomacy, not governance. Bird flu pandemics, the Arab Spring revolutions, the financial crisis, terrorist networks and the spreading of cyber-crime are all manifestations of our ever more connected world. They all illustrate that the current pace of technological change, particularly in the area of ICT, is outstripping our capacity to manage it.

Our society is data-rich, but lacks the conceptual and technological tools to handle it. (Credit: worradirek/Shutterstock.)

The inter-connectedness of global phenomena, and in particular the roles of interactions between individuals, groups and institutions, give a new perspective to events that could look superficially like more of the same. For example, the fall of long-standing, dictatorial regimes in Tunisia, Egypt and Libya was unlike the dissolution of the Soviet Union, not least in terms of its bottom-up impetus. Alleged triggers of the 'Arab Spring', whether they be escalating food prices in North Africa or the self-immolation of a Tunisian street vendor in protest at official harassment, must be seen as catalysts that unleashed rather than created the phenomenon. While the importance of social networking media in these uprisings (which some have called Twitter revolutions) remains open to debate, the issue is not so much whether they 'caused' the revolutions but that their existence – and the concomitant potential for mobilizing a young, educated demographic – can alter the way things happen in North Africa, the Middle East and beyond. Similarly, while economic crashes have always been with us, the financial crisis that began in 2008 was evidently a product of the interconnections – strong ones, yet poorly known – within the institutions that instigated it. The crisis was partly about risk hidden so deeply as to cause paralytic fear; it was also about instruments too complicated for users to understand, and about legal and financial systems labyrinthine enough to permit deception, selfishness and mendacity to thrive.

The Arab Spring of 2011: the product of a complex, deeply interconnected social system. (Credit: MOHPhoto/Shutterstock.)

What is qualitatively new about these events is the crucial role of interdependence and interaction and the almost instantaneous transmission of information through social, economic and political networks. That novelty does not by itself explain why they happened, much less help us to identify solutions or ameliorate the unwelcome consequences. But it points to an unavoidable truth: the world has changed, and it is not going to change back.

We are, for one thing, now living in a world that is data-rich, but with much of the important information highly dispersed so that it can be brought to light only by a smart process of aggregating and sifting. Intelligence may need to rely increasingly not on a few 'hard facts' but on diffuse 'sensing' of mood and opinion: on patterns normally invisible among the noise, such as the epidemiological data unearthed from Google searches by GoogleFluTrends.

Many political analysts today consider that the major challenges in the future will be examples of *discontinuous* change: not gradual shifts in the balance of power or the organization of societies and cultures, but sudden, perhaps catastrophic transformations. Such changes are extremely hard to predict, in terms not just of their magnitude, onset and occurrence but of their very nature – we don't know exactly *what* is going to break.

All this is uncharted territory for politicians, and they do not know how to navigate it. That makes for a dangerous situation, because if political leaders feel compelled to improvise solutions that fail entirely to acknowledge the nature of the problem, they stand a good chance of making things worse. As Lee C. Bollinger, president of Columbia University in New York, has said, "The forces affecting societies around the world are powerful and novel… Too many policy failures are fundamentally failures of knowledge."

This is why politicians and decision makers need new concepts and tools if they are not to lose the capacity to govern, to manage economies, to create stable societies, to keep the world worth living in. And they will need to learn the key lesson of the management of complex, interacting systems: solutions cannot be imposed, but must be coaxed out of the dynamic system itself. There is an analogy with earthquakes, which may never be exactly predictable, but might possibly be managed by mapping out in great detail the accumulating strains that give rise to them, and perhaps inducing controlled, small-scale release of pent-up energy (for example, by injecting groundwater into fault systems). This approach, rather than top-down imposition of laws and structures, might be the way to handle 'social earthquakes' too.

It is sometimes said that by their very nature no one can be expected to foresee radical departures from the

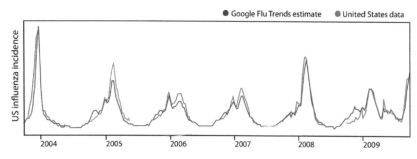

Patterns in the number of searches for influenza-related topics worldwide turn out to closely track flu outbreaks recorded by disease-monitoring centres, with the advantage that the data are available almost instantaneously. See http://www.google.org/flutrends/.

previous status quo. Yet social and political discontinuities are rarely if ever random in that sense, even if there is a certain arbitrary character to their immediate triggers. In the complex systems familiar to natural scientists from the physical and biological sciences, discontinuities don't reflect profound changes in the governing forces but instead derive from the interactions and feedbacks between the component parts. And they are not necessarily unpredictable: sometimes there are precursor signs, and sometimes we can foresee the circumstances in which they will occur, or at least in which they will be more likely to do so.

The notion of 'complex systems' is relatively new in the social sciences. But natural scientists have studied these systems with much success for several decades now. This book argues that the time is ripe – indeed, the need is urgent – to approach the social sciences from this perspective. It calls for a collaboration between natural and social scientists between, for example, computer scientists, physicists, mathematicians, biologists, technologists, psychologists, economists, sociologists, urban planners, political scientists, philosophers, historians and artists – to build a new picture of human social behaviour and its consequences. This is an immense task, but it is already beginning. It is one we can no longer afford to neglect.

Is Society Predictable?

The idea that the social sciences can usefully employ concepts developed in the natural sciences is not new. It was evident at the very origin of modern political philosophy. In the seventeenth century, Thomas Hobbes based his theory of the state on the laws of motion recently deduced by Galileo, in particular the principle of inertia. The ascendancy of the mechanistic view of the natural world, for which the paradigm was Isaac Newton's gravitational model of the cosmos, gave rise in the eighteenth century to a belief that social behaviour also follows rigorous laws that can be expressed and understood along similar mechanistic lines. Adam Smith's notion of an 'invisible hand' that creates a stable and efficient economy from the self-interested behaviour of its many actors already embodied the image of social self-organization that required no over-arching guidance or authority. The operation of this invisible hand was deemed to be as dependable as the law of gravity, provided that the state did not interfere: a central tenet of the belief that markets must be free if they are to be efficient, which many economists and politicians still hold to some degree today.

And in the nineteenth century the cohesion of society as a collective result of the actions of its multitude of members was considered in statistical terms: what mattered was not the capriciousness of individual actions and choices, but the predictable averages. This image both influenced and was influenced by the evolving physical theories of matter envisaged as a vast collection of atoms and molecules: the ideas that gave rise to the twentieth-century science of *statistical physics*. Just as the random, unpredictable movements of individual particles in a gas produce, en masse, the wholly reliable and mathematically simple 'gas laws' that relate its pressure, temperature and volume, so might society show predictable and regular behaviour when viewed as a whole. Thus, early sociology was largely constructed according to an unspoken faith that there was a kind of 'physics of society'.

What is Complexity?

In retrospect, this idea remains valid but it often drew on the wrong analogies. Society does not run along the same predictable, 'clockwork' lines as the Newtonian universe. It is closer to the kind of complex systems that typically preoccupy statistical physicists today: avalanches and granular flows, flocks of birds and fish, networks of interaction in neurology, cell biology and technology. These systems differ from simple gases in that the component particles or agents interact strongly with one another, affecting and responding to one another's behaviour. That is true even for a non-living system like a pile of sand: tumbling grains can strike other grains, setting off cascades that can produce avalanches of all sizes, which are difficult to predict individually but which have characteristic statistical patterns.

This means that societies are more like the communities and ecosystems studied by biologists: food chains, ant and bee colonies, predators and their prey. At one level that seems hardly surprising, for what are societies but communities of a particular species of animal? But what is striking is that analogies between the group behaviour in these cases exist despite the supposedly much greater psychological and cultural sophistication of humans. Some features, such as collective movements and modes of organization, seem rather insensitive to the fine details of how individuals interact, and are determined by the very fact of those interactions, along with the *shape* of the networks they define. That's why descriptions of the resulting behaviour remain accessible to the kinds of theories of complex systems that physicists have developed. They do not necessarily need a great deal of biological or psychological realism to capture the essence of the emergent phenomena.

Thus, on the macroscopic level, social and economic systems have some features that seem to be similar to properties of certain physical or biological systems. For example, they tend to develop hierarchical organization. In social systems, individuals form groups, which establish organizations, companies, parties and so forth. These

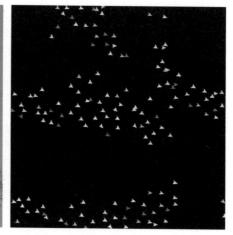

Flocking of birds and fish (left) can be mimicked in a computer model (right) in which each individual reacts only to the motions of its near neighbours. (Credits: (left) Flickr photo by Jef Poskanzer, for free use under Creative Commons licence; (right) prepared using the NetLogo free software, http://ccl.northwestern.edu/netlogo.)

in turn make up states, which might develop alliances and build broader communities of states such as the United States, NATO or the European Union.

The field often called complexity science has evolved to describe systems of this sort. Definitions vary, but there is a general consensus that a complex system is one made up of many components (which might or might not be identical) that interact strongly with one another. When these components are autonomous entities that can make decisions – representing animals, people, institutions and so forth – they are often called agents.

That it has recently become possible (and popular) to study and model systems of this sort is due to several factors. One is the maturity of the discipline of statistical physics, which over the course of the twentieth century developed the theoretical tools and concepts needed to describe and explain the behaviour of increasingly complicated phenomena. But perhaps the most important factor has been the explosion in computer power over the past several decades, which has made it possible to conduct simulations of complex systems in cases where the traditional approach of writing down and solving mathematical equations is intractable.

There are some key concepts that have arisen from these studies, which we will encounter repeatedly in this book. Perhaps the most important is that *complex systems can display ordered, regular types of behaviour*. The apparent complexity of their fundamental nature, with many components interacting with many others, does not necessarily lead to chaos and unpredictability. Rather, there are commonly *emergent collective modes of behaviour*, such as the coherent motions seen in swarms of birds or fish. Here 'emergence' refers to the fact that it is usually impossible to predict this organized collective behaviour by considering the details of the system: by looking at the rules of interaction between the constituent agents. The only way to find out what will emerge is to look and see: for example, to run a computer model. This ability of complex systems to adopt orderly patterns of behaviour is often called *self-organization*: it is not imposed from above, for example by agents all following a leader, but arises spontaneously from the bottom up.

These self-organized modes of behaviour typically appear suddenly – that's to say, a very small change in the forces or properties governing the system, such as the density of particles, can induce an abrupt, profound *global* change in their behaviour. That's something familiar from the early days of statistical physics, when physicists began to understand how it is that substances freeze or melt. These too are sudden changes that happen everywhere: just a tiny drop in temperature below zero degrees is enough to switch water from the liquid state to ice. It looks as if all the water molecules have somehow decided to stop moving at the same time. That's because freezing is a collective property that depends on the interactions between molecules. Freezing and melting are examples of what scientists call *phase transitions*. Models of social behaviour often also exhibit jumps that look like phase transitions – indeed, these are precisely the 'discontinuous changes' mentioned above. They happen at certain *thresholds* in the magnitude of the influences driving the system's behaviour – the density of traffic flow, say, or the proportions of different types of agent.

One of the most important characteristics of many real-world complex systems, especially those in the social sciences, is that they never settle down to a steady, unchanging state. Unlike, say, a block of ice, they are not in equilibrium: they are *non-equilibrium* systems. These are the most challenging systems for statistical physics to describe, and it is only relatively recently that this has become possible. Yet many if not most real-world complex systems are like this. The weather system is one example: patterns recur, often with some predictability, and yet there will never be a state of unchanging weather everywhere.

What makes a system out of equilibrium is a constant input of energy or matter. Weather is never in equilibrium because it is fed by solar energy, which creates gradients of warmth and cold across the planet, driving movements of air and water. Human social systems are also perpetually fuelled in some way – literally so in the case of traffic flow. Lack of equilibrium does not preclude *dynamic* stability, where modes of behaviour or organization remain steady: traffic, for example, can keep flowing at a constant rate under certain conditions.

Commonly, however, non-equilibrium systems experience fluctuations and variations of many sorts and scales. The popular idealized model for this is the pile of sand. With no fresh input of material and energy, it settles into a static state. But if grains are continually dropped from above, the slopes are constantly growing and steepening, only to be relieved by avalanches of grains that might set just a few grains tumbling or might disrupt the whole slope. Each of these events is a *cascade*, where rolling grains collide with and dislodge others, which in turn collide with others, and so on. Cascades are very common in complex social and technological systems: they are manifested in panic selling in markets, or propagating breakdowns in power grids, or the epidemic spread of contagious diseases.

Fluctuations, phase transitions and cascades can make complex systems hard to predict, and still harder to control. But this isn't impossible, once we recognize that complex systems can't usually be *forced* to behave in a certain way by top-down measures. Instead, they must be guided towards one of the modes of behaviour available to them by 'bottom-up' control: by tweaking the conditions or the rules of interaction. It's like guiding the course of a river: you have to work with the flow, or it will just rearrange the banks.

For complex social systems this consideration carries an important message for governance. It does not imply that political interventions are doomed to fail, but just that they must sometimes take other forms from those often advanced today: ones that facilitate the emergence of desirable, self-organized modes of behaviour. Such interventions must happen at a deep level, and with scope for adaptation and flexibility. And they must acknowledge which states of the system are stable, and which are not. None of this is to deny the value of some state-led, top-down regulation – but when that is applied, we must recognize that the consequences may be non-intuitive and difficult to predict.

Despite the shared characteristics of physical and social/economic complex systems, we should not lose sight of the important differences. In social systems, for example, the number of variables involved is typically much larger, the 'rules of the game' may change over time, and the timescales of these changes may overlap. Besides, when we are dealing with humans rather than inanimate particles, we have to consider the technical, financial, ethical and cultural dimensions (which also change over time), as well as the potential for changing behaviour merely by observing or predicting it. Human behaviour involves (among other things) memory, anticipation, emotion, creativity, and intention.

For such reasons, social systems are the most complex systems we know, and are certainly more complex than physical systems. It is scarcely surprising, then, that many social scientists are skeptical about the value of mathematical models. But while the challenges are greater than some natural scientists appreciate, that is no reason for pessimism. The dramatic progress in this field over the past two decades or so gives reason to believe that social complexity is not impossibly complicated. We already have reason to think that many of the qualitative features and behaviours familiar from experience with physical systems remain evident in social ones – and indeed, it would be rather surprising if they did not. This book offers a brief, selective survey of what we have learnt so far – and where the next steps might take us.

Modeling Complex Systems

In the following chapters I describe some of the models that have been devised to study and explain social behaviour. We must then ask: how do we know if they are any good?

In the natural sciences there is, in principle, a clear procedure for answering that. The predictions of the model are tested against experimental results, and the degree to which they match is a fair indication of the degree to which the model is a valid description. If two rival models vie to explain the same phenomenon, that one is preferred which offers the best match to observations.

But in social sciences it's rarely so simple. Testing a model against observations in the real world is often hard enough in the first place, for various reasons. It is extremely difficult – practically, financially and ethically – to conduct experiments on human social systems. And even when this is possible, the number of parameters or variables describing the system is commonly very large, and some may be hard or impossible to quantify. How, for example, does one measure trust? Surveys of opinions are a common tool in social science, but are notoriously tricky to interpret or calibrate. And whereas in the physical sciences the usual experimental approach is to vary one parameter while holding others constant, this might not be possible in social-science experiments.

The problem goes beyond these practical obstacles to experiment and validation. It is not unusual for one model or theory to appear to be supported by experience in one situation, but others in other circumstances. An example is the effect of incentives on productivity in economic theory: incentives seem to work in some cases but not in others.

This situation does not necessarily imply that the theories are inadequate or ambiguous, but may be simply a reflection of what the social world is like. Not only do outcomes often depend on a host of different contingencies, but sometimes there may be too much variability in the system – too sensitive a dependence on random factors – for outcomes to be repeatable. The converse is also true: a particular social phenomenon might be equally well 'explained' by two different models based on quite divergent assumptions.

In such cases, we need not despair of the value of models or theories. Rather, it might be necessary to accept that a particular phenomenon has no unique explanation, no 'best' model that accounts for it. In such cases, there might instead be a need for several complementary and overlapping models, some of which work well sometimes or for some aspects of a problem and others in other cases. This goes against the grain for many natural scientists, although in fact the situation is not unprecedented even for them: predictions of climate change, for example, draw on many different models, which include and exclude different aspects of the 'living Earth' system, and which have various strengths and weaknesses. The predictions are an amalgam of the results of all these models, expressed as a range of possibilities with an estimate of uncertainties and often with an acknowledgement of some outliers that exceed the limits of most model outcomes. A 'majority view' might not be philosophically very satisfactory (who is to say that the outlier might not include some vital factor that the other models neglect?), but it seems to work well enough in practice. It is very likely that a science of social complexity will need to embrace this position of 'pluralistic modeling'.

What this Book is for

The aim of this book is to show that it is possible and productive to try to understand social systems as complex systems, and in many cases to design and direct them with that in mind. That approach has already been found to work in some cases: to offer explanations for social phenomena, and to suggest solutions to social problems, where other, more conventional theories and approaches

have failed. The examples chosen here, both in terms of general topics and specific case studies within them, are by no means comprehensive, but are intended as an illustrative sample of what has been achieved.

As well as demonstrating the general validity of this approach, the book has a more specific agenda. It argues that the complex-systems view of social sciences has now matured sufficiently for it to be possible, desirable and perhaps essential to attempt a grander objective: to integrate these efforts into a unified scheme for studying, understanding and ultimately planning and predicting the world we have made. Such a scheme would not constitute a single 'model of everything', but rather, would allow society and its interactions with the physical environment to be explored through a combination of a suite of realistic models and large-scale data collection and analysis. It is a vision that should now be possible by mobilizing and coupling many different research communities, and it is one that might enable us to find new and effective solutions to major global problems that are impending or already with us, such as conflict, disease, financial instability, environmental despoliation and poverty, while avoiding unintended policy consequences. It could give us the foresight to anticipate and ameliorate crises, and at least to begin tackling some of the most intractable problems of the twenty-first century. The final section of the book outlines a project with these objectives.

Further Reading

D. Helbing, *Social Self-Organization*. Springer, Berlin, 2012.

M. Buchanan, *The Social Atom*. Bloomsbury, London, 2007.

P. Ball, *Critical Mass*. Heinemann, London, 2004.

N. Johnson, *Two's Company, Three is Complexity*. Oneworld Publications, London, 2007.

J. Miller & S. E. Page, *Complex Adaptive Systems*. Princeton University Press, Princeton, 2007.

J. M. Epstein & R. Axtell, *Building Artificial Societies*. Brookings Institution Press, Washington DC, 1996.

F. Schweitzer (ed.), *Self-Organization of Complex Structures: From Individual to Collective Dynamics*. Gordon & Breach, London, 1997.

D. Helbing, 'Pluralistic modeling of complex systems', *Science & Culture* **76**, 315–329 (2010).

Per Bak, C. Tang & K. Wiesenfeld, 'Self-organized criticality: an explanation of 1/f noise', *Phys. Rev. Lett.* **59**, 381–384 (1987).

T. Vicsek, A. Czirok, E. Ben-Jacob, I. Cohen & O Shochet, 'Novel type of phase transition in a system of self-driven particles', *Phys. Rev. Lett.* **75**, 1226–1229 (1995).

J. Ginsberg, M. H. Mohebbi, R. S. Patel, L. Brammer, M. S. Smolinski & L. Brilliant, 'Detecting influenza epidemics using search engine query data', *Nature* **457**, 1012–1014 (2009).

On the Road: Predicting Traffic

Traffic is a problem, and it's going to get worse. Drivers in Los Angeles can expect to spend about 56 hours a year sitting in jams, while every day traffic jams block around 7,500 km of roads in Europe. In Germany alone in 2012, the year's total jam length amounted to 450,000 km, equal to the circumference of the Earth plus the distance between Earth and Moon. But that's mild compared to the situation developing in China, where in the summer of 2010 there was a single jam on the Beijing-Tibet highway stretching for 60 miles and lasting nine days.

Because traffic is one of the simplest and best studied of complex human social systems, it can already be well understood and even predicted. Feedback between real-time monitoring of traffic flows and modelling on computers provides a demonstrated capacity to ease the problems of congestion. What's more, traffic modelling is suggesting new ways of planning road systems so that they are able to accommodate more vehicles with fewer jams. Better design of road networks, junctions and intersections, along with smart management of traffic signals and regulations and the use of automated driver-assistance technology, can help to relieve the pressures that are currently threatening to overwhelm traffic systems across the globe.

This is much more than a matter of reducing inconvenience for road users. Improving traffic flows by treating them as a complex system could also lead to improved road safety and fewer fatalities, would reduce pollution, and would save economies millions from lost working time and inefficiencies of transportation. In the US, traffic delays are said to cost nearly $ 100 billion each year (a figure that has tripled in three decades), and waste around 10 billion litres of fuel.

Beyond this, traffic flow offers perhaps the perfect metaphor for and illustration of many of the general principles underpinning a complex-systems view of society. In traffic we see how the interactions between many notionally 'free-thinking' agents can give rise to robust and inexorable collective modes of behaviour, along with abrupt switches between those modes in response to small influences – a small increase in traffic density, or a minor local disturbance on the highway, say, can plunge moving traffic into a jam. Traffic also illustrates how complex systems like this might be more effectively managed not by top-down control but by faith in the ability of the agents collectively to organize themselves into optimal patterns of movement and behaviour, if the conditions are conducive. In this way, traffic represents a fairly simple case study for validating the general methods and objectives of social complexity science.

Flow States

When we are driving along a road, we are perhaps at our most predictable. No one suggests that our usual autonomy, free will, irrationality or impulsiveness is lost the moment we sit behind the wheel, and yet in general we are prepared to subjugate all this psychological sophistication to the simple demands of the road: we behave as though our only wish is to travel along a linear route from one place to another at a speed that suits us, while avoiding collisions with other vehicles and – on the whole – not flouting legal constraints. In other words, we are placed in a situation in which our behaviour can be described by a few rather simple rules.

This simplicity supplies the basis of modern traffic modelling. Typically, a traffic model will ascribe to each vehicle a preferred 'open-road' velocity, to which it will accelerate if the conditions allow it. But each vehicle will slow down in order to maintain a minimum separation from the one in front. The precise details of this regime of acceleration and braking may differ from one model to another. In particular, it seems that more realistic behaviour is found if it's assumed that drivers also seek a comfortable ride, without jerky speeding up and down. But the basic principles are common to all. It remains then for these simulated drivers to be let loose on a highway system, and to see what emerges.

Traffic models suppose that vehicles move along predefined highways according to simple rules, chosen largely to avoid collisions. (Credit: Courtesy of Argonne National Laboratory - Transportation Research and Analysis Computing Center.)

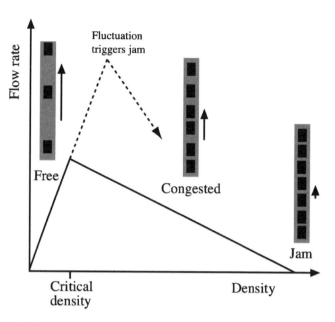

In simple agent-based models of traffic flow, the flow rate (number of vehicles passing a point per hour) increases steadily with traffic density until it reaches a critical value, at which point there may be a switch to a congested state. Then, the flow rate drops as density increases, until finally a more or less stationary jam develops (solid line). But apparently 'free' flow can persist above the critical traffic density (dashed line) unless it is disturbed by some random fluctuation.

Early studies using very simple agent-based models of this sort revealed one of the key considerations in traffic flow: changes aren't necessarily gradual. You might naively imagine that as the volume of traffic on a road increases, it will gradually get more and more congested. Even if that were so, what it would mean for the overall flow rate (how many vehicles pass a certain point on the road per hour) takes a bit of thought. The more vehicles there are on a certain section of road (the greater the traffic density), the more will pass along it even if their speed remains the same. So the flow rate increases as the density increases. But there comes a point at which the vehicles start to interact: they are so close together that they have to adjust their speed to avoid danger. Then any increase in density is offset by a decrease in average speed, so the flow rate levels off. As the density gets higher, the speed might need to drop to a crawl and eventually to a standstill: the flow rate then drops towards zero.

Yet this switch from increasing to decreasing flow rate as the traffic density increases doesn't happen smoothly. Computer-simulation models imply that there is a critical density above which the flow rate stops increasing and abruptly starts to fall: a 'capacity drop'. This sort of threshold effect – a sudden change in the overall behaviour of the system triggered by only a small change in the governing parameters, here the traffic density – is typical of complex systems. It is one of the features that can make them hard to predict: such behaviour is said to be 'nonlinear', meaning that the magnitude of an effect doesn't necessarily follow in proportion to its cause. Precisely the same kind of nonlinearity is familiar in the behaviour of many-particle systems in physics and chemistry, where, as noted in the Introduction, it can give rise to phase transitions such as freezing.

We can in fact draw an analogy between these physical states of matter and the behaviour of traffic. Free-flow traffic, in which each vehicle can move independently of the others at low traffic density, could be considered analogous to a gas, in which the particles barely interact. But if the density of a gas is increased, there comes a point where it condenses abruptly to a liquid, in which the particles do interact but nevertheless remain in motion. This is rather like a 'congested' state of traffic: all the vehicles keep moving, perhaps even reasonably rapidly, and yet they are aware of and constrained by one another's presence. At a still greater density, the liquid freezes to a solid, just as the traffic may 'freeze' into a stationary jam. That this too is an abrupt transition is something all drivers recognize: you may be driving along steadily in heavy traffic, and then within the space of just a few vehicle lengths you are suddenly at a standstill.

Simple traffic models reveal still more. For one thing, the switch from free to congested flow can be postponed beyond the traffic density at which it 'should' happen. In other words, even above the critical density the flow rate can go on increasing with density: the traffic is able to 'pretend' that it is still free. This, however, is a precarious state. As long as all drivers keep moving fast, all is well. But if one individual loses nerve or concentration and brakes too sharply, this tiny fluctuation can trigger a switch to the congested state. In fact, in such a situation a fluctuation can plunge the traffic straight into a jam, from which it is impossible to emerge back into this 'pseudo-free' flow. This makes the traffic flow not only sensitive to small perturbations but also dependent on its own history: it might depend not only on the traffic density but also on how it got to be the way it was. This history- or path-dependence is another common characteristic of complex social phenomena, and it too is familiar from physical systems: water can be cooled below freezing point without actually freezing (that is, it can be supercooled) if this is done carefully so as to eliminate the chances of tiny crystallites of ice growing and 'seeding' the freezing-up of the whole system.

All Together Now

Chance fluctuations, driver over-reactions or overtaking manoeuvres can trigger a change in the flow mode to create the well-known phenomenon of phantom jams: traffic jams that seem to have no cause, and which apparently congeal 'out of nowhere' on busy roads. Congestion may also be triggered by physical disturbances to the flow such as intersections where a side road joins or leaves the freeway, or bottlenecks caused by lane closures. Traffic models have shown that in those situations there are several more possible flow modes than just free, congested and jams.

For example, jams at the point of disruption might grow and ease repeatedly as time progresses. Or a jam might move slowly upstream; or there could be a series of these moving waves of congestion, so that vehicles repeatedly enter and escape from them. It's possible to plot out a 'map' of these various traffic states, called a phase diagram, as a function of the traffic flow along the main highway and that entering at the junction. All of the states seen in these models have been spotted in real traffic. The boundaries between the different states are usually sharp: they are separated by threshold values of traffic flow density. This exemplifies another common characteristic of complex social systems: they tend to have distinct modes of behaviour under different conditions *even when the basic rules governing the interactions of the agents remain unchanged*. Moreover, switches between these modes are abrupt rather than gradual. A big part of the challenge of understanding social complexity lies with mapping out the landscape of possible behaviours. That can help us to understand why, say, changing the underlying rules (such as speed limits) might have a big effect in some situations but none in others, depending on whether or not we are close to one of these boundaries between modes. It might also help to avoid vain attempts to engineer a particular mode of behaviour under conditions where this is simply not stable. The value of models like this is not just that they allow us to predict what will happen under a specific set of circumstances, but that they can offer a global view of the landscape of possible outcomes.

Aside from conventional approaches to engineering the flow of traffic – lane and overtaking restrictions, say, or opening and closing parts of a road network – there are now new technological possibilities for altering the basic rules of vehicle interaction in useful ways. Automobile

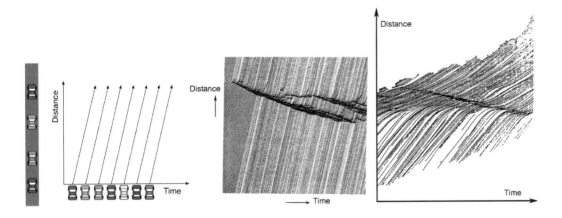

Phantom jams. In a computer model, cars move down a road at a steady speed if their progress is unimpeded (left). But if the traffic is sufficiently dense, a single small and transient disturbance can trigger jams (middle: jams are the dark bands) which move upstream against the direction of driving and develop complex forms, such as splitting into a series of knots of congestion. Jams like this are commonly seen in real traffic data (right). (Credits: (middle) from K. Nagel & M. Paczuski, *Phys. Rev. E* **51**, 2909 (1995); (right) from J. Treiterer et al., 'Investigations and Measurement of Traffic Dynamics', Appendix IV to the Final Report EES 202-2, Ohio State University, Columbus, 1965.)

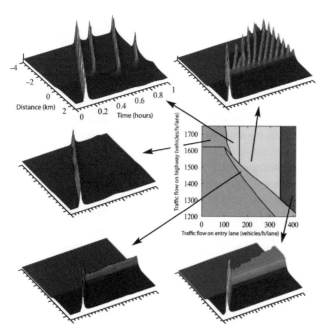

An example of a 'traffic phase diagram' (middle right), showing the different flow modes as a function of the traffic densities on a main road and an inflowing side road. The contour plots show representative flow patterns for some of these states, where high points indicate congestion. (Credit: Courtesy of Dirk Helbing, ETH Zurich.)

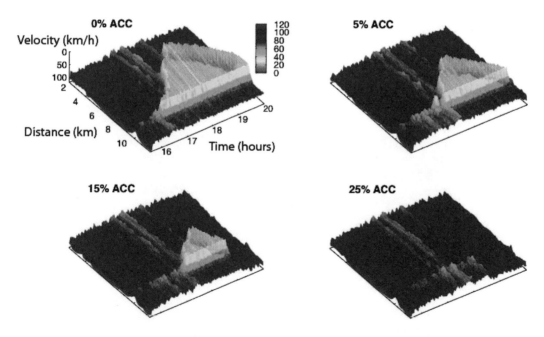

How 'adaptive cruise control' (ACC), an automated driver-assistance system that optimizes the acceleration and deceleration of vehicles to the prevailing traffic conditions, can dissolve a jam. With no ACC (top left), a jam (orange/yellow) develops and spreads upstream of a junction at which new traffic enters a freeway, in this computer simulation with rush-hour traffic densities. As the proportion of vehicles with ACC increases, the jam shrinks, and if 1 in 4 vehicles have ACC (bottom right), it vanishes. (Credit: Courtesy of Dirk Helbing, ETH Zurich.)

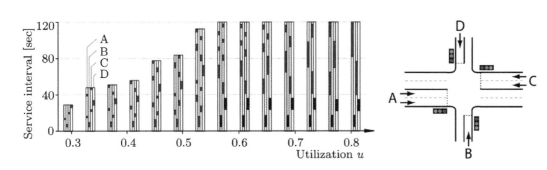

Adaptive cycles of traffic-light operation (off = solid bars) for the four-way, multi-lane intersection shown on the right. The optimal sequence changes as the overall incoming traffic density (represented here by the 'utilization' u) increases. (Credit: Courtesy of Dirk Helbing, ETH Zurich.)

manufacturers are developing 'driver assistance' systems that automate aspects of driving, not just to lighten the load on the human driver but to improve safety and reliability. For example, radar and laser sensors can detect the proximity of other vehicles or obstacles to avoid collisions, alert drivers who begin to drift out of a lane, help with lane changing, and allow for smooth hill descent without the driver having to apply brakes. The improvements offered by such systems – for example, making braking manoeuvres smoother – can help to avoid jam-triggering fluctuations. It has been estimated that some jams in heavy traffic can be 'evaporated' if just one-fifths of the vehicles have driver-assistance systems that enable them to respond optimally to changes in traffic flow. Here, then, new information and communication technologies can help not only to understand and predict complex social dynamics but literally to steer them towards more desirable states.

Traffic Lights Can Organize Themselves

Improving traffic flow is not so much a matter of imposing particular behaviours as of creating the conditions under which the traffic can spontaneously organize itself in the most efficient manner. That philosophy doesn't always sit easily with managers and planners, but it is looking increasingly to be the best way to approach all manner of social phenomena: to relinquish top-down control in favour of a faith in the bottom-up capacity of complex systems to find their own efficient modes of behaviour, given the opportunity.

This is illustrated by a reconsideration of how to coordinate traffic lights. The normal approach is to synchronize a periodic sequence of on-off times for a series of lights at a cluster of intersections. But there is no reason to suppose either that the best sequence is strictly periodic or that it has to remain the same regardless of the traffic conditions. It turns out to be better to allow each individual traffic light to respond adaptively to the flow conditions at any moment.

The notion of allowing traffic signals individual autonomy could sound like a recipe for disaster – why should there be any guarantee that what is 'best' for one intersection will also suit what is happening at the others? However, if each signal is supplied with information not only about the traffic at that particular junction it but also about the traffic coming from neighbouring junctions, this sharing of information can enable the system as a whole to find more effective, flexible solutions at any moment than are available from an insistence on regimented periodicity.

The idea is that traffic sensors placed a little before an intersection feed information about the incoming flow to each individual light-controlling system. This makes it possible to calculate the expected delays, and corresponding 'traffic pressures', at different parts of the network. Priority for green signals is then given to those parts experiencing the greatest pressures. In this way, the traffic itself controls the lights, rather than vice versa. Chance fluctuations, such as temporary lulls in the traffic on some routes, can be exploited to relieve congestion elsewhere. The behaviour that emerges can in fact look surprisingly synchronized, for example in the appearance of 'waves' of green lights that travel through the network.

Simulations of traffic flow on a network where autonomous lights are coordinated in this way show that overall average delays can be reduced by 30–40 % relative to today's state-of-the-art conventional control methods, and that the travel times for individual vehicles through the network actually become more predictable.

A simulation study of a real-world urban network – 13 intersections in the busy city centre of Dresden, complicated by tram lines and pedestrian crossings – has shown that self-organized traffic lights can significantly reduce the waiting times for all the modes of transport, including pedestrians. The same principle can be applied to other traffic-control measures such as speed limits: allowing them to adapt to the prevailing flow conditions can reduce congestion and delays.

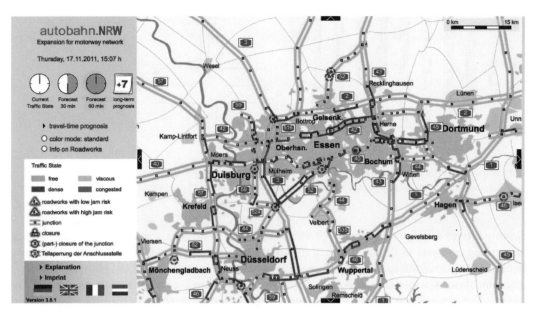

A snapshot of the traffic in the Duisburg-Dortmund region of Germany predicted one hour ahead, based on simulations using a model called OLSIM. See http://www.autobahn.nrw.de.

Theory Into Practice

The key principle here again serves as a metaphor for the management of social complexity more generally. What is needed for efficient solutions is bottom-up autonomy of agents, informed by feedback from other parts of the system, and guided by real-time data that samples the instantaneous behaviour in the real world and uses it as the basis for anticipating probabilistically what is going to happen in the near-term: an x percent chance of *this*, a y percent chance of *that*. In other words, we don't impose solutions, but rather, adjust the rules governing the interaction of agents so that robust and effective modes of behaviour can emerge.

Models that can predict traffic flow on a real road network, based on measurements of traffic at a few key locations, should be valuable for planning journeys and for alerting road authorities to potential congestion problems before they arise, thereby potentially averting them. Schemes like this are already being implemented in European and American cities and urban areas to allow real-time prediction of road use. One, called the Transportation Analysis and Simulation System (TRANSIMS), has been used to plan road networks in Dallas, Texas, and is now being expanded to give a broader vision of urban transportation networks in other cities (see page 47). Another one uses models like those described here to forecast traffic on the autobahn system in the North Rhine-Westphalia region, encompassing 2,250 km of highway. Schemes like this should not only make traffic flow more freely but could also make them safer, reduce pollution, and make the design of transportation networks more rational, flexible and sustainable.

Further Reading

D. Helbing, 'Traffic and related self-driven many-particle systems', *Rev. Mod. Phys.* **73**, 1067–1141 (2001).

D. Helbing & M. Treiber, 'Jams, waves, and clusters', *Science* **282**, 2001–2003 (1998).

D. Helbing & K. Nagel, 'The physics of traffic and regional development', *Contemp. Phys.* **45**, 405–426 (2004).

A. Kesting, M. Treiber, M. Schönhof & D. Helbing, 'Adaptive cruise control design for active congestion avoidance', *Transport. Res. C* **16**, 668–683 (2008).

S. Lämmer & D. Helbing, 'Self-control of traffic lights and vehicle flows in urban road networks', *J. Stat. Mech.* P04019 (2008).

B. S. Kerner & H. Rehborn, 'Experimental properties of phase transitions in traffic flow', *Phys. Rev. Lett.* **79**, 4030–4033 (1997).

A. Pottmeier, R. Chrobok, S. Hafstein, F. Mazur, M. Schreckenberg, 'OLSIM: Up-to-date traffic information on the web', Proc. 3rd IASTED International Conference: Communications, Internet, and Information Technology, November 22–24, 2004, St. Thomas, US Virgin Islands.

Every Move You Make: Patterns of Crowd Movement

Walking from here to there doesn't seem like the most complex choice we face in our lives. Don't we, like light beams or the proverbial crow, just take the most direct route, proceeding at a pace that suits us? But exploring and navigating our environment on foot is rarely so simple. What if there are obstacles in our path? Robotics engineers have long realized that it is no mean feat to find a compromise between directness and smoothness of trajectory, with no abrupt changes of direction. What if the terrain varies – some paved, some grassy or muddy? Trickiest of all, what if the objects in our way are themselves moving, if they are other pedestrians headed somewhere else? How crowds move around open spaces is a sophisticated process of collective negotiation that depends vitally on how we interact with one another.

Getting about in a crowd is a complex affair. (Credit: Courtesy of Michael Schreckenberg, University of Duisburg.)

Extraordinarily (self-)organized movement can arise from many individual decisions uncoordinated by any leader, as is clear from the flocking of birds, the swarming of fish or bees, or the foraging of ants. These collective motions in the natural world have become almost emblematic in complexity science of the way that coherent, coordinated modes of behaviour can *emerge* from simple behavioural rules – we now know that flocking can arise merely from each bird responding in simple ways to what its near-neighbours do. And because they are not imposed 'from above', these group motions can be highly adaptive to changing circumstances – for example, to the arrival of a predator.

But human group motions typically lack such elegance, in part because the intentions of the group are not usually so unanimous. Shoppers in a busy city plaza have a hundred agendas and destinations. Some are in a hurry; some enjoy meandering; some are in family groups, perhaps with small children; some are on roller-blades or have motion-impairing disabilities. But all have at least one thing in common: they don't want to bump into anyone. And that, it seems, is largely all it takes for a crowd to find what are often surprisingly efficient ways of self-organizing.

Knowing how people move around public spaces should enable architects and urban planners to design them more efficiently and safely. Designers of shops and galleries might want their floor plans to allow and indeed encourage everyone to go everywhere. Architects want to know where to put doorways so that they will be used most efficiently; likewise for park designers deciding where the footpaths should go. 'Dead spots' in urban spaces may deter traders and encourage crime and decay.

There are more urgent reasons to know how crowds move around. Rock concerts, sports events and religious festivals have all been blighted by disasters in which crowd congestion has caused crushes and fatalities. Avoiding such hazards is partly a question of identifying bottlenecks and flashpoints, but it may also depend on understanding how and why crowd movements can themselves change qualitatively – what makes a congested group of people switch from a frustrating but orderly shuffle to a panicked crush? What types of collective motion are safe, and which can presage catastrophe?

Walking on Computers

Social scientists have long spoken of 'social forces' that govern behaviour. Some of the earliest 'complexity' models of crowd dynamics took this image rather literally, positing forces of attraction and repulsion between individuals that determine the routes they take. It's not that we really do exert such mutual forces – rather, we act as though this is so. In particular, we avoid collisions as if a repulsive force holds us back from bumping into each other.

On this basis, it is possible to develop pedestrian models that are rather like the traffic models described in the previous chapter, in which individuals move through space towards their destination at a velocity that will decrease or change direction if necessary to avoid collision. This minimalist picture of pedestrians, when simulated on the computer, produces complex emergent behaviour that looks not only surprisingly reminiscent of the real world but also somehow imbued with intelligence, anticipation, even good manners.

One of the simplest scenarios sets the walkers travelling in both directions down a narrow corridor. If the crowd density is high, one might expect to see chaos and congestion. But the 'model crowd' can avoid this by adopting collective orderliness. The pedestrians arrange themselves into counterflowing streams in which individuals track each other's path. This makes sense: following the person in front of you means you are far less likely to collide with someone coming the other way. But there is no ingredient in the model that insists on 'following' behaviour – it's simply what appears once the rules are allowed to unfold.

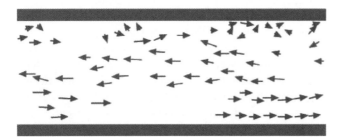

Streams in a computer model of pedestrians in a corridor. (Credit: Courtesy of Dirk Helbing, ETH Zurich.)

This streaming can be enhanced by adding obstacles such as pillars to divide the flow, even without specifying which side the walkers should pass on. Similarly, model pedestrians coming to a multi-way intersection may self-organize themselves intermittently into an efficient circulating motion, as if at a roundabout – and this pattern is enhanced if there is an obstacle such as a pillar at the centre of the intersection. At doorways, walkers moving in opposite directions will organize themselves into what almost resembles a traffic flow controlled by lights, with alternating bursts of egress through the opening in one direction and the other – a relatively efficient way of minimizing congestion. Each group stands back from time to time to make way for walkers coming in the other direction, giving an illusory appearance of cooperation simply as a result of the desire for collision-avoidance.

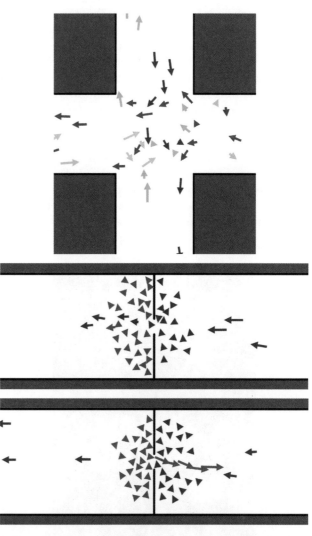

Snapshots of model 'walkers' negotiating a crossroads (top) and a doorway (bottom, showing two snapshots where 'red' and 'blue' capture the doorway). (Credit: Courtesy of Dirk Helbing, ETH Zurich.)

Spontaneous lane formation is also found in some animal group movements, such as those of foraging ants that

Trails spontaneously trodden over grassy spaces have a curved, 'organic' feel to them (left). This is what emerges in a pedestrian model too (right). (Credit: Courtesy of Dirk Helbing, ETH Zurich.)

lay down pheromone trails to attract others. This helps to avoid collisions between ants leaving and returning to the nest. Humans don't follow scent trails in the same way, but there are other situations that induce us to follow in each other's footsteps even when we don't see one another. On open grassy spaces such as public parks, we have a tendency to walk where others have previously worn the grass away, both because it requires slightly less effort and because of a psychological impulse to 'stick to the path'. The path taken by a walker over an open space is therefore a compromise between this trail-following behaviour and the wish to take the most direct route. In this way spontaneous paths are created and reinforced, leaving an imprint of the collective consensus about the preferred routes. The trails must be sustained by use: if old paths are abandoned, grass eventually obscures them.

Pedestrian models can account for the non-intuitive geometry of these human trails. Over time, the most direct trails between the points of entry and exit are replaced by curved routes, with islands of grass isolated in the middle of intersections. This information could guide planners and architects to put paths where they will be used, rather than trying to impose a pattern that might be undermined or ignored.

Panic

Self-organized crowds can thus find collective modes of movement that ease the flow and enable people to move past each other with the minimum of congestion and discomfort. But not all crowds are so civilized. In crowd disasters, panic can lead people to push over others and trample them. Riots, building fires, and crushes at sports stadiums and rock concerts have in the past claimed many lives as a result of the uncontrolled movements of crowds gripped by panic.

Pedestrian models show a switch from orderly motion to what looks like panic as the desire to move fast overwhelms the urge to avoid collisions. A crowd like this can become jammed if all the individuals try to pass quickly through a single doorway, for example to escape from a burning room. As walkers press in against one another in the dense throng in front of the door, they can become locked into arch-shaped lines, unable to move forward. Such locked arches of particles can cause salt to get stuck in the salt cellar even though each grain is small enough to pass through the hole. The result is counterintuitive: as everyone tries to move faster, the crowd exits through a doorway more slowly than it would if everyone kept to a moderate pace. Very high pressures may build up in a jammed crowd: in real crushes they have been large enough to bend steel barriers and knock over walls. Mice trying to escape from a flooded chamber show just the kind of 'escape panic' exhibited by these computer model, with flow through the exit occurring in short, erratic bursts.

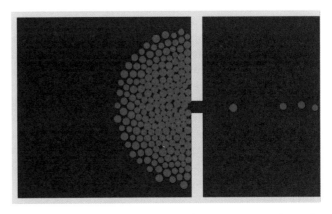

A simulated crowd trying to exit rapidly from a room can get jammed in a 'panic state', in which individuals escape only slowly and sporadically. (Credit: Courtesy of Dirk Helbing, ETH Zurich.)

One of the key findings of these panic simulations is that exits are generally not used efficiently: some become

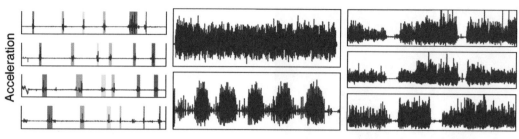

Patterns of acceleration in individuals (each represented by one of the horizontal traces). On the left, people queueing are revealed by sporadic bursts of movement. In the middle, the top trace shows a person walking freely while the lower trace shows a person moving down a clogged corridor. On the right, the top two people can be seen to be moving together, while the bottom person is not. (Credit: from D. Roggen et al., Networks and Heterogeneous Media **6**, 521–544 (2011).)

jammed while others are neglected. This can result from herding behaviour: individuals tend to follow others rather than locating exits themselves. There is an optimal level of crowd-following behaviour which allows exits to be used efficiently: too little herding means that people fail to benefit from others having found a way out, whereas too much means that a few exits get jammed while others remain barely used. In any event, room-emptying calculations based on average equal use of all exits may not be adequate for anticipating how well they serve in an emergency.

Better Rules, Better Data

Although pedestrian models based on attractive and repulsive forces seem to do a fair job of reproducing some real-world behaviour, they can scarcely reflect the rules that govern the movements of real people. Models that are more rooted in cognitive psychology substitute these fictitious forces for a set of so-called heuristics: rules of thumb that mimic more closely how we make decisions. For example, it seems that walkers adjust their speed and direction by estimating when a collision would occur if they don't take evasive action, and by seeking the least deviation from the direct route to their destination. That's to say, instead of being repelled by their neighbours, pedestrians seek the most convenient free path among them. Using heuristic rules like this, pedestrian models can provide rather accurate predictions of the precise trajectories that people take past obstacles and other walkers.

While these models of crowd behaviour display many realistic features in a qualitative manner, they also highlight the fact that we don't actually know very much about precisely how people *do* move around. Planning for crowd management and safety has long relied on collecting information about, for example, how heavily used particular exits or routes are, but this very coarse-grained information says little about the paths of individuals, let alone about what governs them. Such purely empirical findings aren't much help for determining how, why and when the collective behaviour of a crowd changes – when, for example, orderly motions break up into chaotic, turbulent ones. Neither is it straightforward to know or predict the effects of altering the underlying constraints on crowd movement (changing access routes, say).

In short, studies of pedestrian movements need more data. New technologies make it increasingly easy to gather them. For example, safe and unobtrusive radiofrequency tags are already used to track the whereabouts of individuals who are considered potentially at risk. These can be used to gather information not only about movements but face-to-face encounters, as was recently done in a study on school children to understand how such contacts might contribute to the transmission of infectious disease. In fact we already carry tracking devices most of the time in the form of mobile phones, which can be used to deduce a lot about how we navigate our personal worlds (see page 29).

Sensors worn on the body can reveal not only where we are but how we are moving: accelerometers, for example (which are also contained in most mobile phones), register changes in speed or direction. In one recent study, accelerometers were used to track changes in the walking speeds of many people at a public event in Malta. Distinctive patterns could be seen for different types of collective behaviour. Queueing was characterized by short bursts of activity separated by periods of stasis, while motion down a crowded corridor led to periodic clogging. By comparing individual data sets, one can deduce which people are walking together in groups, and it might be possible to identify signals in individual motion that presage and herald changes in the overall dynamics of a crowd. This is just one example of how information and communications technology can, at very little expense, enrich our picture of social complexity as manifested in daily life.

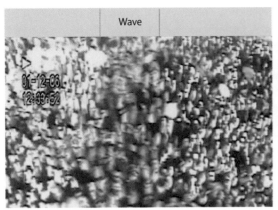

Crowd movements at the 2006 Hajj in Mina, revealed in video recordings that average images over 1–2 seconds so that people in motion look blurred. As the crowd became denser, stop-and-go waves appeared (left), and, for still denser crowds, earthquake-like turbulent motion could be seen (right). (Credit: Courtesy of Dirk Helbing, ETH Zurich.)

Planning Public Spaces

Town planners and architects have too often planned public spaces according to how they think people ought to move around them, rather than how people would naturally like to do so. As a result, studies have shown that pedestrians in some urban developments such as housing estates seem to traverse the space more or less randomly: at the statistical level, they look lost. These spaces have clearly not been designed with people in mind. The effect is not just that human intentions are thwarted, but that our movements and decisions might in turn be directed by our environment. By the same token, modeling of visitor trajectories in the Tate Britain art gallery in London has shown that over 70 percent of the differences in visitor numbers to each room could be accounted for purely in terms of the ground-plan layout, without any reference to what the rooms actually contained. As Winston Churchill once said, "We make our buildings and afterwards they make us. They regulate the course of our lives."

Computer models of pedestrian motion should make it possible for our buildings and spaces to be shaped by us rather than shaping us. They are now being integrated into architectural design schemes. For example, the London-based company Space Syntax has used them to help plan the pedestrianization of the busy tourist destination of Trafalgar Square in central London, and is currently working on urban-design problems in cities ranging from London to Jeddah and Beijing.

Pedestrian simulations have been used to help design crowd-management and safety measures for large public events. In one study, a model simulating crowd movements in the Notting Hill carnival in west London was able to identify flashpoints where there was a high potential for dangerous congestion, so that stewarding and crowd-control measures could be concentrated there.

Modelling of crowds served an even more urgent need in a study of the annual pilgrimage of Muslims to Mecca in Saudi Arabia. This event, called the Hajj, draws more than three million pilgrims, and hundreds of lives have been lost in the past as a result of overcrowding and stampedes. In 1990 over a thousand people were trampled to death in a pedestrian tunnel leading to the so-called Jamarat Bridge in the nearby town of Mina, where a symbolic ritual stoning of pillars (the jamarat) takes place. Despite a major redesign of this area, the stoning in Mina has been a flashpoint for several disasters: fatalities have occurred on six occasions since 1994. In January 2006 over 300 pilgrims died, and many more were injured.

At that event, video cameras installed to monitor the movements of the crowd happened to record that, as the crowds on the Jamarat Bridge became denser, the steady flow changed to stop-and-go waves like those in heavy traffic. And as the crowd density increased still further, people were pushed around by an earthquake-like state denoted 'crowd turbulence', in which uncontrolled shock waves threatened to make people lose their balance and fall to the ground.

The onset of this hazardous turbulence happens when a parameter equal to the crowd density multiplied by the variation in speed of motion exceeds a certain threshold value. Real-time monitoring and analysis of video data on dense crowds can give advance warning of when this highly dangerous state is about to develop, so that interventions might relieve the pressure and avoid the onset of a catastrophic accident.

With the guidance of crowd modelling, a new route between the pilgrim camps and the Jamarat Bridge in Mina was devised for the 2007 event, with time schedules arranged to limit and distribute the flow of pilgrims. The redesigned route was successful: despite the even higher number of pilgrims than previous years, the Hajj passed without incident. This was an indication that models of pedestrian motions can have a real and immediate impact on our daily lives, helping to make the built environment more conducive, more accessible, and safer.

Further Reading

D. Helbing, J. Keltsch & P. Molnár, 'Modelling the evolution of human trail systems', *Nature* **388**, 47–50 (1997).

D. Helbing & P. Molnár, 'Social force model for pedestrian dynamics', *Phys. Rev. E* **51**, 4282–4286 (1995).

D. Helbing, I. Farkas & T. Vicsek, 'Simulating dynamical features of escape panic', *Nature* **407**, 487–490 (2000).

M. Batty, 'Predicting where we walk', *Nature* **388**, 19–20 (1997).

D. Helbing, A. Johansson & H. Z. Al-Abideen, 'Dynamics of crowd disasters: an empirical study', *Phys. Rev. E* **75**, 046109 (2007).

M. Moussaïd, D. Helbing & G. Theraulaz, 'How simple rules determine pedestrian behavior and crowd disasters', *Proc. Natl. Acad. Sci. USA* **108**, 6884–6888 (2011).

D. Roggen, M. Wirz, G. Tröster & D. Helbing, 'Recognition of crowd behavior from mobile sensors with pattern analysis and graph clustering methods', *Networks and Heterogeneous Media* **6**, 521–544 (2011).

M. Batty, *Cities and Complexity: Understanding Cities with Cellular Automata, Agent-Based Models and Fractals*. MIT Press, Cambridge, Ma., 2005.

http://www.spacesyntax.com/

Making Your Mind Up: Norms and Decisions

3

While the causes of the Arab Spring of 2011 are still disputed, no one imagines that the popular uprisings in Tunisia, Egypt and Libya happened one after the other by pure coincidence. Even before President Mubarak of Egypt was forced to relinquish his rule in February, there was talk of a 'domino effect' and 'copycat' behaviour across the North African nations, leading political leaders in Algeria, Morocco and Libya to lower the prices of essential goods in an attempt to avoid spreading of this 'contagious' unrest. Such knock-on effects of social disturbance and conflict were nothing new. For example, the civil war in Rwanda in the 1990s spread into the neighbouring Congo, at the cost of millions of lives, and also affected Burundi and Uganda.

At much the same time as the Arab Spring was demonstrating that the mass behaviour of one group can influence that of another, a similar phenomenon was witnessed in the United Kingdom. Rioting and looting in north London in response to the shooting and death of a man being arrested by the police triggered similar outbreaks of violence in many parts of England, apparently without a specific motive and sometimes in areas that were not conspicuously deprived and which had never before experienced serious unrest of this sort.

One of the most baffling and disturbing aspects of the English riots, aside from their unexpected and unprecedented occurrence in itself, was the varied demographic profile of the rioters. While politicians lamented the existence of pockets of society lacking any restraining norms of civility, those convicted of the disturbances in fact came from many different social strata. And yet despite lacking any orchestrating agency, the riots ceased as abruptly as they had begun. The phenomenon was reminiscent of episodes of spontaneous 'mass hysteria' recounted by historians.

This capacity for sociopathic behaviour to spread like a disease has long been remarked upon, and feared – most notably, it has been invoked to explain the rise of genocidal regimes such as those of Nazi Germany, which seem to involve collusion between extremist leaders and the populace. But by the same token, benevolent or neutral behaviour can be propagated too. For better or worse, mobile phone technology has altered the norms and the etiquette of public conduct in ways that no one consciously consented to. Behaviour deemed normal three decades ago, such as racism and homophobia, is widely considered unacceptable in many nations today. (Conversely, one suspects that the tide could turn back just as quickly, as witnessed by the rise of far-right movements in some parts of Europe and Russia.)

In the light of such developments, the mechanisms and dynamics of collective opinion formation are arguably among the most important questions facing the social sciences today. They can determine whether products and ideas succeed or fail. They set the bounds of how we may or may not behave in public and in private. They go

The English riots of 2011: an example of spontaneous group decision-making mediated by new information technologies? (Credit: adapted from the blog http://vikkilittlemore.wordpress.com/ .)

to the very heart of what it means to live in a democracy: how, for example, we negotiate as a society the balance between personal freedoms and collective responsibility. It may be the ebb and flow of collective opinion formation, rather than the aggregate of independent decisions, that determines who our leaders are and what powers and responsibilities they are able to exercise. It seems likely that major challenges facing the planet, such as the threat of climate change, will be determined not so much by laws, legislation and international negotiation as by whether or not a sea change in public opinion can be triggered.

All in This Together?

Much of the debate around multiculturalism in Europe has centred on the issue of whether societies with a plurality of views on religion, authority, politics, personal responsibility and state intervention can operate as coherent entities. Regardless of where we feel the balance must lie, it is generally agreed that societies are not tenable without shared norms. At the apparently trivial level (although the genesis of any custom is far from trivial), we must all agree on which side of the road to drive. More contentiously, we must establish norms, or at least bounds of acceptability, that govern choices of clothing, taxation levels, personal honesty, privacy and much else. These norms are necessary not only for social stability but also because we could barely function as individuals without them. Norms exist so that we do not need to decide afresh all the time how we should behave, dress, or speak.

In his 1978 book *Micromotives and Macrobehaviour*, the economist (and later Nobel laureate) Thomas Schelling pointed out how decision-making is an interactive social process, the outcome of which is not always predictable from an inspection of individual beliefs or behaviours. Schelling's book pioneered the vision of society as a complex interactive system, although these concepts were then still in gestation. Schelling presented his book as an exploration of how the 'invisible hand' that Adam Smith evoked to explain the self-organization of the economy might operate also in social life. Yet he included in his analysis the very factor that economists have tended (to their cost) to ignore: interaction. As Schelling put it,

> We usually have to look at the system of interaction between individuals and their environment, that is, between individuals and other individuals or between individuals and the collectivity. And sometimes the results are surprising. Sometimes they are not easily guessed. Sometimes the analysis is difficult. Sometimes it is inconclusive. But even inconclusive analysis can warn against jumping to conclusions about individual intentions from observations of aggregates, or jumping to conclusions about the behavior of aggregates from what one knows or can guess about individual intentions.

Schelling's classic illustration of how interaction can create non-intuitive outcomes in collective decision-making concerned the issue of social segregation. At that time, several US cities had experienced so-called 'white flight', in which affluent white people left the inner city for the suburbs, leaving behind deprived ghettos populated by racial minorities. This rather extreme racial segregation seemed indicative of a high degree of racial intolerance. But was it really?

Schelling described a model in which two kinds of agent – call them red and white – move about on a checkerboard grid according to simple rules based on the colour of their neighbours. An agent is selected at random and, if more than one-third of the eight nearest neighbours are found to be of the other colour, it will move to one of the free spaces on the grid where this will not be the case. After each agent had moved on average just three or four times, an initially highly mixed population became strongly segregated into islands of a single agent type. Schelling's model has since been explored in various different incarnations, and can be seen to be formally analogous to the physical separation of liquids, such as oil/water or molten metal alloys.

On the one hand, this model suggested that it takes only a small degree of individual prejudice – indeed,

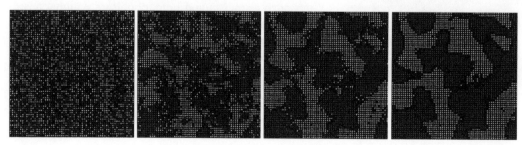

Segregation of two populations in Schelling's model happens very quickly as time progresses (from left to right). Image kindly supplied by Michael Batty. (Credit: Courtesy of Michael Batty, University College London.)

what some might regard as a natural human desire to be in a majority – to create rather extreme segregation. The collective effect of many marginal decisions is stronger than each might individually seem to imply. On the other hand, the implication seems to be that segregation may be inevitable, or at least rather hard to suppress.

In other words, the political implications are open to debate. Must we accept segregation (by race, culture, religion, class…) as a fact of life? Or can we use models like this to find ways of rendering it less likely? Must segregation in fact be seen as a 'bad thing'? Evidently the growth of ethnic districts in cities can contribute to their diversity and to the avoidance of bland homogeneity, as well as offering mutual support and identity to their residents. But segregation can also lead to unrest and mutual distrust of communities.

Regardless of these open and difficult questions, Schelling's model makes an important general point about interactive social systems: individual behaviour cannot necessarily be extrapolated to infer group behaviour. More specifically, when individuals are offered the opportunity to express choices by aligning with one community or another, there is a strong likelihood that segregation will quickly develop. While Schelling's original model seems particularly prone to exhibit strong segregation, related agent-based models that allow the agents to exploit the benefits of diversity still struggle to find stable mixed states: there is always a pull towards segregation.

This message is not always acknowledged within political ideologies that promote choice. For example, one recent study of choice and segregation in US schools concluded that "Programs that allow the unfettered movement of children across schools will exacerbate existing race and class-based segregation in traditional, local neighborhood schools that children leave, further deteriorating the educational conditions faced by the most disadvantaged students." However, other studies in the UK offer mixed evidence for whether school choice policies have increased segregation. If anything, the evidence so far underlines the need for these social models to become considerably more sophisticated: not just to offer agents a simple binary choice (move/stay), but to acknowledge the many factors that influence decisions of this sort, such as the interplay between in-group and out-group determinants such as race, class and wealth, and the often complex cost/benefit considerations of different choices. In other words, understanding this important social phenomenon has only just begun. However, the problems associated with segregation – such as the formation of poverty traps, so-called 'sink estates' and 'no go' areas, the tensions associated with immigration, and the increasing disparity in some countries between private and public provision of education, health and security – all show that this is a vitally important area for future research.

Peer Pressure

In situations like these, we express our choices by "voting with our feet" – by moving house or moving school, say. But social norms and codes, and indeed the outcomes of any democratic voting process, tend to arise in a more interactive, negotiated manner. In particular, we often attempt to persuade others to make the same choice as us.

The study of opinion formation and the emergence of social consensus has witnessed an explosion of activity in the past two decades. Models in which the decision-making agents are arrayed on a grid and exert 'forces' that encourage neighbours to adopt the same behavioural 'orientation', like spinning compass needles, are formally very similar to microscopic models of magnetism, in which each magnetic atom in a crystal influences those around it. Under some conditions the system can settle into complete consensus – all the 'needles' point in the same direction, as they do in a piece of magnetized iron. But the system can also break up into patches with orientations that are uniform within the patch but different from those of other patches – a situation long known in magnetic materials. And if there is too much randomness influencing individual orientations, even local consensus becomes impossible, just as thermal jiggling will destroy the magnetic ordering of iron above a certain temperature. Such simplified 'opinion models' are largely well understood now, but run the risk of becoming more a matter of formalized physics than a description of what happens in a real society.

A simplified representation of a voting model, in which agents arranged on a grid are like magnetic needles that try to force their neighbours to point in the same direction.

Studies of opinion formation commonly now try to inject more real-world relevance. What if the agents are arranged not on a grid but in a branching network that better reflects the shapes of our own social networks of peers? As Chapter 5 shows, the topological structure of a network of interactions – who is connected to whom – can make a big difference to the collective behaviour that emerges. In this case, a realistic network structure can, relative to a simple grid, boost the number of distinct opinions that survive in the case where no outright consensus emerges.

And what if some agents are more influential than others, or more stubborn in sticking to their views? Under what conditions does a consensus tend towards moderate opinions, and when do extremist views come to dominate? How are opinions altered by overall 'orienting fields', which could mimic the effects of advertising or media bias?

The results of such studies offer some provocative conclusions. For example, extremist groups that are resistant either to altering their own views or to interacting with those with very different views can, if they are not wholly isolated, actually help to maintain a diversity of opinion in the rest of society. (This frustration of consensus could be regarded as either a positive or negative outcome.) In some situations, the presence of extremists can eventually catalyse a switch to an extreme consensus, even after a prolonged phase of near-consensus on a more moderate opinion.

Agent-based approaches like this can be extended to look at how cultural traits spread through a population as individuals tend to copy their neighbours or accede to the traits of the local majority. Must majority-copying lead to a monoculture, a 'Macdonaldization' of diversity? Even for simple models this is a subtle question, which can depend on factors such as how big the overall population is and how rich the cultures are (that is, how many different 'features' they contain that distinguish one culture from other). For example, there seems to be a threshold of 'cultural richness' beyond which a single dominant culture can never colonize the entire population. There are obvious implications here for issues such as commercial globalization and the survival of minority languages. So far, most of the models remain rather too abstract to draw reliable conclusions about the real world, but they already establish that complex behaviour can emerge from simple rules and that it's unwise to rely too heavily on intuition to predict how the normative behaviour of many agents will evolve.

The influence of public and peer opinion on our own choices has long been recognized by social scientists, but has been difficult to demonstrate or quantify. An ingenious social-science experiment accomplished this in 2006. The researchers enlisted over 14,000 volunteers who could listen online to songs recorded by 48 unknown rock bands, and download ones they liked. Social influence on these choices was studied by presenting the information to the participants in different ways. Some were simply shown a list or grid of the songs in random order. The number of subsequent downloads was then assumed to be a measure of the 'objective quality' of the songs – not in any formal sense, but in terms of how this was perceived by the study group. Other subsets of the group were provided with information about the song's popularity, in terms of the number of times it had been already downloaded. This information was supplied in two different ways: either as a bare number attached to a random listing of the songs, or as a list ranked in order of popularity. The latter makes the 'social information' much more transparent, and thus increases the strength of the social influence.

This influence was found to have a marked effect on the choices made. In particular, the stronger the social interaction, the more 'inequality' there was in the outcome: popular songs were more popular, and unpopular ones less so. In other words, the choices reinforced one another when they were known to other members of the group. Moreover, when there was this feedback about the choices of others, 'quality' (as measured by the group that made decisions independently) became a less reliable indicator of a song's 'success', and increasingly so as the social interaction got stronger. While 'bad' songs never did particularly well, and 'good' ones rarely did poorly, all things seemed possible in between: social interaction made it more likely that mediocre songs could become runaway successes. In other words, there was greater unpredictability of outcomes, since small, random differences in individual judgement could become amplified by social feedback into major differences in the group rankings.

Reputation and Trust

Findings like this take on considerable significance once we recognize that many of our choices rely ever more on social interaction and feedback rather than on a traditional deference to central authority. As consumers, we might be swayed by online recommender systems that attempt to correlate our own past choices with those of other users. Rather than buying from a few trusted major stores or suppliers, we shop amongst a huge variety of providers on eBay and other online markets. We receive a constant stream of suggestions, statements and influences from Twitter streams. Our sources of information and news are attaining a bewildering diversity in terms of cultural and political orientation, standards of rigour and accountability, and so on. Increasingly, we navigate these potentially hazardous waters by relying on feedback that advises us of the reputation of our sources and thereby determines our degree of trust. For providers, the active construction of reputation is vital. Influence is not only reflected in but also determined by the quantity and the opinions of followers and users. One of the revelations of today's networked world is that this mechanism of trust-building, which aggregates the evaluations of peers rather than relying on the monolithic, institutional status of traditional stores, brands, critics and experts, can engender a self-organized system of trade, exchange, debate and discourse that allows individuals to trust one another

Customers Who Bought This Item Also Bought

The Strategy of Conflict by Tc Schelling ★★★★★ (1) £16.46

The Evolution of Cooperation by Robert Axelrod ★★★★★ (9) £9.99

The Winner's Curse: Paradoxes and Anomal... by Richard H. Thaler ★★★★☆ (4) £19.90

Capitalism and Freedom by Milton Friedman ★★★★½ (13) £7.70

Our choices rely increasingly on those of others, conveyed to us by automated systems like the 'recommender systems' of online providers.

(or not) without any previous first-hand encounters. We are all, in a sense, mining the collective experience in order to make judgements and choices.

How reputation and trust are maintained seems likely to be one of the most significant and fertile areas for exploration through a complex-systems approach. One of the lessons of the financial crisis is that trust has always been the invisible lubricant of the financial system – and that, when it evaporates, the consequences are catastrophic. The same might already be true also for our commercial and political systems. So we had better understand how, in these webs of advice and persuasion, such things as influence, consensus and power arise and can be manipulated.

Further Reading

T. C. Schelling, *Micromotives and Macrobehavior*. W. W. Norton, New York, 1978.

W. Weidlich, *Sociodynamics: A Systematic Approach to Modelling the Social Sciences*. Harwood, Academic, Amsterdam, 2000.

D. Stauffer, 'Opinion dynamics and sociophysics', in *Encyclopedia of Complexity & System Science*, ed. R. A. Meyers, 6380–6388. Springer, Heidelberg, 2009.

S. Galam, 'Opinion dynamics, minority spreading and heterogeneous beliefs', in *Econophysics and Sociophysics* ed. B. K. Chakrabarti, A. Chakraborti, A. Chatterjee, pp.367–391. Wiley, Berlin, 2006.

G. Weisbuch, 'Social opinion dynamics', in *Econophysics and Sociophysics* ed. B. K. Chakrabarti, A. Chakraborti, A. Chatterjee, pp.339–366. Wiley, Berlin, 2006.

R. Axelrod, 'The dissemination of culture: A model with local convergence and global polarization', *J. Conflict Resolut.* **41**, 203–26 (1997).

F. Gargiulo & A. Mazzoni, 'Can extremism guarantee pluralism?', *J. Artific. Societ. Soc. Simul.* **11**, http://jasss.soc.surrey.ac.uk/11/4/9.html (2008).

G. Deffuant, F./ Amblard, G. Weisbuch & T. Faure, 'How can extremism prevail? A study based on the relative agreement interaction model', *J. Artific. Societ. Soc. Simul.* **5**, http://jasss.soc.surrey.ac.uk/5/4/1.html (2002).

S. Suo & Y. Chen, 'The dynamics of public opinion in complex networks', *J. Artific. Societ. Soc. Simulat.* **11**, http://jasss.soc.surrey.ac.uk/11/4/2.html (2008).

D. J. Watts & P. S. Dodds, 'Networks, influence, and public opinion formation', *J. Consumer Res.* **34**, 441–458 (2007).

M. J. Salganik, P. S. Dodds & D. J. Watts, 'Experimental study of inequality and unpredictability in an artificial cultural market', *Science* **311**, 854–856 (2006).

M. Mäs, A. Flache & D. Helbing, 'Individualization as driving force of clustering phenomena in humans', *PLoS Comput. Biol.* **6**, e1000959 (2010).

Broken Windows: The Spread and Control of Crime

4

Early social statisticians in the nineteenth century were astonished to find that the prevalence of crimes from year to year followed a precise mathematical pattern. Specifically, variations from the long-term average fitted onto a bell curve – the same curve that described variations in births and deaths, or in errors in experimental measurements. This means that small deviations from the average happen more often than large ones, and that very large deviations tend to be vanishingly rare. How could it be that these acts of crime, committed with free will, obeyed such regularity? To some, such as the influential Belgian mathematician Adolphe Quetelet, this meant that crimes must be somehow compelled by a higher force – not by individual choices, but by "the customs of the people", and that they must therefore be regarded as an inevitability: a "budget that is paid with frightening regularity".

We are now generally less fatalistic about crime. The vast literature of criminology is predicated on the idea that crime has specific causes, whether psychological or social, and that these can, at least to some extent, be understood – which should then suggest ways in which crime might be reduced, even if not eliminated. However, this research has produced no consensus. Is crime committed after a rational cost-benefit analysis of the gains and risks, as some economists have suggested, or is it done in the heat of the moment? There is evidently a link to socioeconomic circumstances, but this link is not straightforward – many socially deprived areas have low crime rates, while there are also 'crimes of the wealthy' such as tax avoidance. Do deterrents work? Do they work better if they are harsher? Because there is no simple or single answer to such questions, approaches to crime prevention are often based more on faith and ideology than evidence.

But what about those bell-curve statistics that impressed Quetelet? We now know that there is nothing so surprising in them, for a bell curve is the characteristic signature of random fluctuations in independent events – you find the same thing in statistics of coin tossing, where each toss takes no heed of the last. These statistics suggest that, whatever the reasons people commit crimes, they do so independently of one another.

Yet crime is not always like that. In March 2010 a man stabbed eight children to death in an elementary school in southeast China. Within two months there were four other such child attacks of a similar nature, killing 17 people in total, in different provinces. Although this is an extreme example, the phenomenon of the copycat crime is well attested, and often seems likely to be sparked by wide, sensationalist media coverage of the original event. Crime is often spoken of as an epidemic, with the implication that it involves an element of contagion that causes criminality to spread and wane episodically. Knife crime in Britain and organized drug crime in Latin America, for example, have been described in these terms. This picture implies that, not only can criminality be 'caught' from others, but so can lawfulness.

These aspects of crime – the evident but complicated dependence on a range of social 'drivers', and the feedback effects between behaviours – make it a complex social phenomenon as much as a question of individual psychology. Even in people who have identifiable mental-health problems that might give them a greater propensity towards criminal activity, such as violence, it seems that social factors, including the behaviour of other people, can determine whether such traits are manifested or not.

With prison populations on the point of overload in some developed countries, methods of policing under the spotlight, and the possibilities for both criminal activity and security monitoring being multiplied by information and communication technology, understanding crime has become one of the most urgent problems for governments and societies. A complex-systems approach cannot guarantee to provide all the answers, but it undoubtedly shows the promise to provide a broader perspective that allows criminality to be embedded in a more inclusive landscape, as an aspect of social decision-making that depends, among other things, on the establishment of norms, changing demographics, multiculturalism, economic conditions, urban planning and education. This perspective can accommodate the evident fact that crime is not, after all, a random and individual act but a collective social behaviour. As such, it is unlikely that a linear cause-and-effect approach will do much to solve it. Moreover, seeing crime as a complex social system encourages a multi-tiered strategy towards its alleviation: applying local measures to thwart it in the short term, say, while implementing deeper, longer-term social changes to attack the root causes.

On the Streets

Some early models of crime as a complex system have neglected social interaction and focused instead on other factors known to influence the likelihood of offences. It has been clearly shown, for example, that the incidence of burglary is related to characteristics of the built environment: it is more common in terraced houses, say, and tends to be directed at wealthier areas and at students (obviously for different reasons!). And burglars prefer not to travel far from their own homes. These spatial and demographic factors are hard to embody in some theoretical approaches, but agent-based models are well suited to including them. One recent model that used these factors to determine the likelihood of burglaries in the English city of Leeds achieved a reasonably good match to observed data provided that the modelling was conducted at a detailed spatial resolution of about 1 km – an indication that too much aggregation of statistics ('crime rates have increased by 15 percent') may mask important information.

Most crime can be considered to be a social interaction, albeit an unpleasant one, between the criminal and the victim, analogous to the encounters of predators and their prey in the wild. Criminological studies have shown that victim behaviour can be very important in determining the susceptibility to crimes such as household burglary: most obviously, it is more likely when the residents are not at home. For this reason, some models have taken both perpetrators and victims explicitly into account. One criminological theory supposes that crimes occur when routine activities bring a potential offender into contact with a potential victim in a circumstance where the offender calculates that there is sufficient absence of 'guardianship' – police officers or members of the public likely to intervene – to make the crime, and consequent economic gain, worth the risk.

In this picture, patterns of routine activity for both parties become key. Street robberies, for example, are highly sensitive to the patterns of human movements, which in turn are dependent on the spatial structure of the urban environment as discussed in Chapter 2. In one study, movements of three types of agent – law-abiders, who might be either potential victims or 'guardians', police, and potential criminals – was modelled on the street network of Seattle, with demographic and socioeconomic distributions matching the real data for that city. This model aimed to create realistic patterns of movement based on the locations of homes and centres of activities such as retail, recreation and employment. This realistic 'siting' of the simulation turned out to be crucial: street robberies were much more tightly clustered for a 'real' Seattle than for a simple grid street network, reflecting the greater convergence of victims and offenders along certain popular routes. Even if the quantitative predictions of a model like this are open to question, it makes the general point that, at least for robberies and muggings, it is essential to make space an explicit part of the picture. Identifying crime hotspots in an urban space, for example, shows where policing needs to be concentrated. This model also lends support to the theory that crime increases as people spend an increasing amount of time away from their homes.

Model predictions of street robberies in Seattle, showing a concentration on major thoroughfares. (Credit: Courtesy of Elizabeth Groff, Temple University, Philadelphia.)

Including victim behaviour in the above-mentioned model of household burglaries in Leeds also improved its realism: for example, how affluent they are and over what proportion of time their house is occupied. Such detailed information is not usually available for specific houses in a geographical area, but anonymous census information of this sort can at least be used to make the model representative of local characteristics. One of the findings in this case is that location may be a dominant influence on the likelihood of being burgled: unemployed people living near potential burglars were more likely to be victims than affluent individuals living further away, even though the former houses are more often occupied (and probably have less to offer the burglar). The next step for such models is to increase the psychological complexity of the agents: to base decision-making on the kinds of motivations known to be relevant to real offenders, including their awareness of their environment (such as potential target houses for burglary) and their perception of risks. In such models, different agents will develop different 'cognitive spaces' – mental maps of their surroundings and behavioural rules of thumb – that affect their choices.

As with other simulations and models of complex social systems, the value of these studies may lie not so much in identifying what will happen where and when, but in facilitating counterfactual 'virtual experiments' that can explore the general consequences of particular policies and events in ways that cannot be done by real-world experimental testing. We can ask, for example, what is the effect of intensifying or reducing policing, or how socioeconomic changes, shifts in the age demographic, and increased surveillance, affect crime levels.

Bad Influence

The idea that crime breeds more crime has a long pedigree, but it has been very hard to demonstrate. It assumes that we will be more likely to behave in antisocial or irresponsible ways if we see others doing so. One could regard this from the perspective of game theory (see Chapter 8): if everyone around us is acting selfishly and for their own gain, we will be suckers if we don't do the same. By the same token, cooperative 'good' behaviour by our peers not only makes us feel safe enough to act likewise but can create a social pressure that enforces 'goodness' with the threat of ostracism.

While there is surely some truth in this picture, the effect of interaction on social behaviour can also be more indirect and subtle. We seem to react not just to the actions we see around us but to proxies of them imprinted on the environment.

This is the so-called 'broken windows' hypothesis of sociologists James Q. Wilson and George Kelling, which supposes that people are most likely to commit criminal and antisocial acts when they see evidence that others have already done so – for example, when they are in public places that show signs of decay and neglect.

This idea motivated the New York subway system's famous zero-tolerance policy on graffiti in the late 1980s (for which Kelling acted as a consultant), which is credited with improving the safety of the network. The idea is that simply by removing evidence of neglect on the subway, users were prompted to behave better: not just to refrain from painting graffiti, but to be more law-abiding all round.

Some have even argued that the graffiti removal spurred the general reduction in crime in New York City between the 1980s and the 1990s. Others believe the causes lay elsewhere: better policing, say, or an improvement in the economy or an increasing average age in the population. In any event, the crime rate in the city undoubtedly changed profoundly in a relatively short space of time: the numbers of both violent assaults and robberies halved between 1989 and 1999. This transformation certainly has the appearance of an 'epidemic of lawfulness'.

Did the removal of subway graffiti solve New York's crime problem? (Credit: ChameleonsEye/Shutterstock.)

Although its influence on crime rates remains an open question, there is now good evidence that the 'broken windows' effect is real: that criminal and antisocial behaviour is affected by what we infer about the behaviour of others from our environment. This is not merely a copying effect: it is well-known that people drop more litter in a setting that is already litter-strewn, but that doesn't imply that they will indulge other antisocial habits in the same place too. However, experiments in the Netherlands have shown that visual evidence of the violation of one norm of 'good behaviour' does encourage people to violate others. For example, cyclists were significantly more inclined to drop on the ground an advertising flyer attached to the handlebars of their parked cycles when these were located in front of a wall on which graffiti defied a prominent notice that prohibited it, than when the wall was clean. Pedestrians would ignore 'no entry' signs into fenced-off areas when bi-

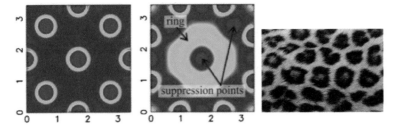

Crime hotspots, here shown in red, can arise in a 'reaction-diffusion' model of the interaction of criminals and victims (left). Under some circumstances, if these are suppressed locally by policing (purple), then the crime might simply be displaced into the surrounding neighbourhood (middle, green ring). These patterns arise in a manner analogous to the biochemical processes thought to produce pigment spots in animal skins (right). (Credits: (left and middle) from Short et al., Proc. Natl. Acad. Sci. USA **107**, 3961–3965 (2010); (right) W.Scott/Shutterstock.)

cycles were left locked to the fence in defiance of signs prohibiting this. And they were twice as likely to steal money from an envelope lodged in a letterbox when the box was defaced with graffiti or when the ground was littered than when it was clean.

These findings show rather dramatically, perhaps even shockingly, that many of us are not either categorically law-abiding and considerate or criminal and selfish, but may display either trait in the face of simple, subconscious cues about behavioural norms. They provide good reason to believe that criminality has a significant dependence on social interaction.

Hotspots

Social feedback offers one explanation for why crime seems to occur geographically in hotspots. This phenomenon is well attested but poorly understood. Sometimes hotspots can be suppressed by increasing police activity in those regions, but sometimes this merely moves the focus of crimes to adjacent areas. Considering the system as one in which 'predators' (offenders) seek 'prey' (victims) while both move around in the available space casts the problem within a framework familiar to ecologists, who have long described such interactions using so-called 'reaction-diffusion' models developed in chemistry, in which spatial patterns can arise as the 'agents' diffuse through space and react with one another. In the case of crime, the 'reaction' – predation of criminals on victims – can be potentially suppressed by an inhibiting agency such as a security measure or a police force. Inhibition in chemical reaction-diffusion schemes can lead to spatial patterning, with different concentrations of the reagents persisting in some regions even while the individual molecules continue to diffuse. Biochemical processes of this type are thought to underlie the formation of pigmentation patterns on animals. Reaction-diffusion systems can be equivalently modelled within an agent-based approach.

A similar process might explain the formation of crime hotspots. In this model, the hotspots turn out to come in two varieties. In one case they are merely aggregates of individual crimes with overlapping spheres of influence. The other sort of hotspot is caused more directly by positive feedback: crime induces more crime. The first sort of hotspot can be eradicated completely by a sufficiently strong inhibiting influence: that is, by locally concentrated policing. But the second sort is harder to eliminate. Focused inhibition can cause the hotspots to break up into smaller spots or rings in the close vicinity. If this picture is an accurate reflection of the world, it suggests that not all hotspots will yield to the same style of policing, but that different strategies might be needed in different situations.

Although models like this have not yet been used to devise or guide law-enforcement strategies, there is growing recognition that a more empirical approach to policing based on detailed crime statistics can be valuable. The Santa Cruz police department in California has been running a trial in which crime data are fed into a computer algorithm which uses statistical inference to predict where crimes will happen next. The algorithm identifies hotspots for offences such as car thefts and house burglaries, not simply by looking at where these have occurred previously but by using an approach analogous to that used to predict the locations of aftershocks after earthquakes: the high-risk areas aren't necessarily ones that have just experienced crime spikes. This approach may bear short-term fruit for law enforcement, but it relies merely on statistical correlations and offers little real insight into why crimes happen. While it demonstrates the value of having more data about crimes, such data might yield a better harvest once they are used to inform models rooted in the movements and behaviours of real people. For while it has been said that predicting crime is like predicting the weather, the weather does not respond to deterrents and incentives. Not all crime is rational, but a rational approach is surely needed to understand it.

Further Reading

L. Liu & J. Eck (eds), *Artificial Crime Analysis Systems: Using Computer Simulations and Geographic Information Systems.* Hershey, Pa., Information Science Reference, 2008.

Special issue on Simulated Experiments in Criminology and Criminal Justice, *J. Exp. Criminol.* **4**(3), 187–333 (2008).

E. R. Groff, 'Simulation for theory testing and experimentation: an example using routine activity theory and street robbery', *J. Quant. Criminol.* **23**, 75–103 (2007).

E. R. Groff, ''Situating' simulation to model human spatio-temporal interactions: an example using crime events', *Transactions in GIS* **11**, 507–530 (2007).

N. Malleson & M. Birkin, 'Towards victim-oriented crime modelling in a social science e-infrastructure', *Phil. Trans. R. Soc. A* **369**, 3353–3371 (2011).

N. Malleson, L. See, A. Evans & A. Heppenstall, 'Implementing comprehensive offender behaviour in a realistic agent-based model of burglary', *Simulation: Trans. Soc. Model. Simul. Int.* doi: 10.1177/0037549710384124 (2011).

N. Malleson, A. Evans & T. Jenkins, 'An agent-based model of burglary', *Envir. Planning B: Planning & Design* **36**, 1103–1123 (2009).

K. Keizer, S. Lindenberg & L. Steg, 'The spreading of disorder', *Science* **322**, 1681–1685 (2008).

M. B. Short, P. J. Brantingham, A. L. Bertozzi & G. E. Tita, 'Dissipation and displacement of hotspots in reaction-diffusion models of crime', *Proc. Natl Acad. Sci. USA* **107**, 3961–3965 (2010).

The Social Web: Networks and Their Failures

5

In 2003 a massive power blackout on the east side of North America affected one third of Canada's population and one in seven people in the USA. The cause was disputed – some sources blamed a lightning strike in the Niagara region, others a fire at a power plant in New York or an alleged breakdown at a nuclear power plant in Pennsylvania. A report in 2004 finally alleged that a computer error caused the shutdown of a generating plant in Ohio when the demand became too high. This put a strain on nearby high-voltage power lines, which failed when they came in contact with overgrown trees. As a result, less than three hours later the problem had cascaded to shut down 256 power plants over the east coast region.

This sensitivity to what seemed like a minor disruption would appear to imply that the electricity network was dangerously fragile. But such huge failures are rare, whereas localized problems are fairly common. The 2003 power cut reveals not so much that the network is delicate, but that its behaviour is extremely hard to predict: some, perhaps most disturbances create only small-scale failures, but others can be catastrophic.

Events like this are significant in themselves for the security and robustness of our societies: had the power failure occurred in winter, the consequences might have been far graver. Physical networks play vital roles in the operation of society, whether they are webs of roads and streets, or telecommunications lines, or supply networks for water and gas. But many other complex social and technological networks exist in a more abstract, intangible realm: the webs of trade and travel that link global air and sea ports, say, or the networks of friends and associates, or those that arise in business and commerce, or which connect ideas and innovations.

None of these networks was designed. They have all grown spontaneously, lacking any blueprint that dictated how the nodes should be joined together. While each link in the electricity grid will have been discussed, planned and executed, no one had responsibility for the overall shape, and until recently no one thought to ask what it looks like, let alone suspected that the answer could have vital implications for the grid's performance. All of these complex social and technological networks are in some sense the product of human actions and intentions, and yet their contours are often unknown. We must explore them as we would aspects of the natural world. And indeed it is now becoming clear that they share many characteristics with networks that exist in nature, such a food webs or the communication between genes and proteins in our cells.

The 2003 American power failure illustrates why it is so important to understand these networks. Research over the past decade or so has revealed not only that their topologies – the patterns of connectivity – are quite different from those that have been assumed in the past, but also that these topologies hold the key to their performance and robustness: how easily the webs can be negotiated and searched, and how sensitive they are to breakage of links. A small effect can have major consequences; conversely, its influence might remain small. As a result, interventions that seek to alter the networks must be considered with great care, since there is a strong potential for unanticipated consequences. As if this were not complicated enough, many networks in society are hierarchical and multi-functional: they are networks of networks. The flow of physical entities – people, materials, energy – along networks might be guided by the flow of information in cyber-networks, and might itself be influenced by other interdependencies such as geographical proximity or the interactions of institutions and authorities. It is precisely in those interdependencies that possible vulnerabilities and catastrophic failures lurk.

Both the mathematical study of networks in an abstract sense, and interest in real social networks, have histories stretching back for at least half a century. Nonetheless, 'network science' can claim to be a genuinely new discipline, having begun in earnest only in the 1990s. It is now one of the most vibrant, diverse and insightful areas of complex-systems research, which has cast a new light on many areas of the social, natural and engineering sciences, from trade and business to epidemiology (the subject of

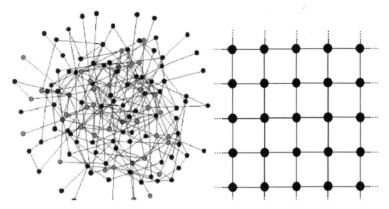

A random network (left) and a grid network (right). The former, but not the latter, typically contains shortcuts between any two points. (Credit: (left) Courtesy of Hawoong Jeong, Korea Advanced Institute of Science and Technology.)

the next chapter), ecology and computer technology. It impinges on all the other topics in this book. As indicated in the Introduction, inter-connectivity is one of the defining and determining features of the modern world.

Small Worlds

One of the most striking and important characteristics of social networks that develop 'organically' through the unplanned addition of new nodes and links – for example, networks of friends and acquaintances, or of professional collaborators, or of links between web pages on the World Wide Web – is that there is almost always a shortcut connecting any two points. As the popular notion puts it, 'it's a small world': we regularly find that new acquaintances already share a friend, or a friend of a friend, with us. This is the message of the famous 'six degrees of separation', a phrase inspired by experiments conducted in the 1960s by the social scientist Stanley Milgram. He sent letters to several people in Omaha, Nebraska, asking them to forward the letters to a stockbroker from Sharon, Massachusetts who worked in Boston. Milgram provided just the target's name, and instructed recipients to send the letter to someone else they knew personally who might be better placed to know the stockbroker. Of the (rather few) letters that reached their intended destination, an average of just six steps was required. Modern experiments using email contacts have verified this finding, although the precise number of 'degrees of separation' is open to debate.

Early attempts to map out networks of contact and friendship tended to assume that they were random, meaning that any new node (individual) joining the web became linked to others at random. (Of course friendships do not exactly form at random, but the assumption was that statistically they look as though they do.) One way of characterizing a network is in terms of how well connected it is – that's to say, how easily it may be traversed from any one node to another. This depends on the average number of links per node. If this number is small, some nodes will remain unconnected or linked only into isolated groups. But if there are many more links than nodes, there is a good chance of finding a route between any two nodes in the network: it is 'fully connected'. For random graphs the distinction between these two situations is rather abrupt: above a threshold value of average connectivity, the largest interconnected component grows in size very rapidly, quickly approaching the size of the entire network.

In that case, there are in general many alternative routes between any two points, and it is very likely that some will involve only a few jumps – there will be shortcuts, where the random linkage of nodes has connected two distant ones. This makes such a random network a 'small world'. Compare this to a grid network like that of the Manhattan street system. There are no shortcuts here: to get from an intersection in Tribeca, say, to one in Harlem, you have to traverse much the same number of links between junctions no matter which route you take.

Small-world social networks are some of the best-explored examples of complex networks, because in some cases all their links have been catalogued. One such is the network of collaborations between movie actors. This has been popularized by the so-called Kevin Bacon Game, in which one must link any named actor to Kevin Bacon in as few steps as possible (see Box, page 25). The average path length in this network is 3.65, which means that on average any actor can be connected to any other in between 3 and 4 links: there are 'three and a half degrees of separation'. The same is true of several other collaborations. For example, the network formed by jazz musicians active between 1912 and 1940 has an average path length of 2.79.

However, just because a network is a small world does not imply that it is a random network. In fact many if not most social networks are not. With hindsight, this isn't

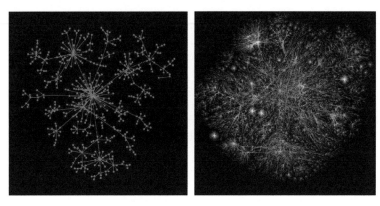

Scale-free networks are non-random small-world networks, which typically have a 'pinched' appearance in which some nodes have a disproportionately high connectivity (left). The Internet has this structure (right). (Credits: (left) prepared using the NetLogo free software, http://ccl.northwestern.edu/netlogo; (right) Courtesy of http://www.opte.org/.)

surprising. The random-network model assumes that any two of your good friends have no greater chance of knowing each other than they have of knowing anyone else in the population. But that's unrealistic: friends tend to form clusters with a strong degree of interconnectivity, with fewer links between clusters. The same is true of other social networks: companies tend to cluster in their mutual dealings, as do collaborating scientists and musicians.

Clustering implies that there's a bias towards forming new links to other nodes already close by: in other words, a bias against making short cuts. Nevertheless, it turns out that a rather high degree of clustering can develop in networks without compromising the shortcuts that a truly random network displays. This is the situation for many real small-world networks.

There are several distinct classes of such networks, but one of the most common is exemplified by the network of hyperlinks between pages on the World Wide Web. One survey of a representative small section suggests that any web page is connected to any other by an average of 19 links – a number that changes only very slowly as the network grows. The topology of this network is revealed by the statistics of connectivity: that is, how many nodes have different numbers of links. This statistical distribution has a form known as a power law: the number of web pages having k connections ($N(k)$) is proportional to the inverse of k raised to some power α: $N(k) \sim 1/k^{\alpha}$. Systems whose structure is described by power laws have the general property that they are scale-free – there is no characteristic size to them. What this means here is that there is no 'typical' number of links to a node of the network. In other words, the connectivity of nodes is highly unequal: some are much better connected than others. The power-law statistics also mean that, compared with a random network, the WWW has a disproportionate number of very highly connected pages. The physical structure of the Internet – the actual links between computers – also has this topology.

Many social networks are scale-free, with power-law distributions of connectivity: for example, email communications, direct flights between airports, trade links between countries, and the movie-actor network. Where does this topological structure come from? Many social networks grow 'organically' by the addition of new nodes. Scale-free networks grow according to the rule that a new node connects to an existing node at random but with a bias: the more links a node already has, the more likely it is to be chosen. So well-connected nodes are likely to become even more so: a feedback effect often called the 'rich get richer' principle or the Matthew Principle, after the Gospel of Matthew: "For unto every one that hath

Six Degress of Kevin Bacon

Here's an example of how to play the Kevin Bacon Game for the case of Elvis Presley. Elvis never appeared in a movie with Bacon himself, but he was in Harum Scarum (1965) with Suzanne Covington, who appeared with Bacon in Beauty Shop (2005). So Elvis has a Bacon Number of 2. The average Bacon Number for all movie actors is about 2.98, which implies that Kevin Bacon is indeed better connected than an average actor. But there nevertheless thousands of actors with comparably small average path lengths, and several hundred are better connected than Bacon. The best is currently Dennis Hopper, with a Bacon Number of 2.80. For details, see http://oracleofbacon.org.

shall be given, and he shall have in abundance". This is basically the principle used in the Google page-ranking scheme: in assessing the connectedness and thus significance of a particular page, more highly connected incoming pages are given a higher weighting.

The bias in the growth rule for scale-free networks means that the connectedness of different nodes does not necessarily reflect real differences in their significance (how 'good' a web page is, say). The positive feedback means that 'fame' artificially inflates some nodes over others. In other words, even if some slight superiority makes some nodes intrinsically more 'attractive', this feedback can blow it out of all proportion – a phenomenon that sounds all too plausible as a reason for the success of some books, music, commercial products, or indeed movie stars.

Clubs and Communities

Small-world networks, then, can be random or scale-free. But these are not the only options, and indeed not all social and technological small-world networks are scale-free. Many power grids, for example, are not, and it is not clear that friendship networks are either. In any event, much of the current work on networks focuses not so much on the overall topology as on the substructure: the hierarchical divisions of nodes into distinct communities or modules. On the Internet the communities of users tend to be defined partly by geography and partly by profession. Similarly, friendship networks might be structured around a neighbourhood or a workplace. This kind of modularity is a reflection of the high degree of clustering characteristic of small worlds. There are now several mathematical techniques for deducing the community structure of complex networks. Extracting this 'buried' information can reveal how we organize our lives and the information we access. For example, a community-finding scheme applied to purchases of books on US politics through the online bookseller Amazon.com showed a division into communities buying only the 'liberal' books or only the 'conservative' ones, suggesting that people tend to read things that reinforce their own views.

One common feature of small-world networks is that highly connected nodes have a greater-than-average chance of being linked to other highly connected nodes, forming what has been dubbed a 'rich club'. There are rich clubs, for example, in the collaboration networks of scientists, movie stars and company directors. Their existence in social, economic and professional networks could have significant implications for the way society functions. Members of a 'rich club' might, for instance, share privileged information that percolates only slowly into the rest of the network.

The shape and form of networks can have a crucial bearing on how it performs its function. For instance, how does the topology affect the ease with which a network can be navigated? A key characteristic of social small-world networks is that they are searchable: we can often find the shortcuts, as Milgram's experiment demonstrated. The origin of this searchability seems to lie with the community structure. If most people belong to several different communities (work, family, friends, sports club and so on), then messages or information can be efficiently routed through the network even though no individual is able to see the whole map: it's sufficient that individuals be aware only of their local links to friends and colleagues a short 'social distance' away. Thus, 'unexpected' shortcuts may be found so long as the network contains many overlapping communities.

Cascade Failures – a Growing Problem

On 4 November 2006, an electricity line was temporarily turned off in Ems, Germany, to enable the passage of a Norwegian ship. This caused a chain reaction that left many parts of Europe without electricity.

On 22 December 2010, Skype initiated a faulty auto-update of its Internet telephone software. This led to a crash and reboot of most Skype 'super-nodes', a crucial part of their distributed system. To make matters worse, the reboot of the super-nodes launched a distributed 'denial of service' attack on the central Skype servers, incapacitating worldwide traffic.

Heavy solar storms disrupt telecommunications from satellites, and have in the past caused failures of some networks. As these networks become increasingly interconnected, the potential for massive failure cascades grows stronger, with the danger that cash machines, sales and customer supplies, computer and communication systems could all fail critically at the same time over large regions.

The current economic crises began with a 'local' problem: a bursting real-estate bubble in the US. The mortgage crises eventually hit building companies and caused the bankruptcy of more than 400 US banks. It continues now to endanger the stability of the European currency and even of the European Union. Several countries, including Greece, Ireland, Portugal, Spain, Italy and the US, are on the verge of bankruptcy. A second crisis seems likely to fuel social unrest, political extremism and increasing crime and violence.

When Systems Fail

Flows, whether of material, information, rumour or disease, on scale-free networks are not easily disrupted. If a few links get broken, the network is in no danger of falling apart or becoming non-navigable, because the many shortcuts means that alternative routes can generally be found. The nodes remain continuously connected even for almost total breakdown of links. In random networks, in contrast, only a few broken links can fragment the network into isolated clusters. In other words, one could say that random networks are shattered by failures while scale-free networks are slowly deflated. This robustness to node or link failures is a fortunate and unplanned result of the way scale-free networks like the Internet have grown.

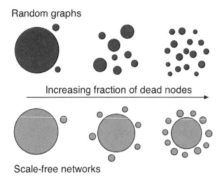

Random removal of nodes has different effects on random and scale-free networks. The former are fragmented by many such breakdowns, while the latter tend to remain largely connected, deflating only slowly even when a large proportion of nodes have 'gone dead'.

But the resilience of scale-free networks relies on the fact that they contain some very highly connected hubs. If, rather than nodes failing at random, the hubs are destroyed selectively, it's a different story. Deactivating just 18 percent of the most highly connected nodes of the Internet would break it into tiny pieces. That is a crucial insight for defending against the intentional disruption of computer networks, known as cyberwarfare – an objective that has now become a major concern for national and international defence and security agencies.

Moreover, in contrast to their resistance to node breakdowns, small-world networks seem especially prone to failure cascades like that which caused the 2003 US blackout (see Box, page 26) It is possible to design networks that don't have this weakness, but they are generally not small worlds, having rather long average path lengths between nodes. This implies that there is a tradeoff between resilience to cascade failure and ease of navigation.

Yet cascade failures and vulnerability to targeted attack may not be inevitable even in small-world networks. It might, for example, be possible to tailor them for 'failure response' strategies such as reinforcing the most highly connected nodes. To implement such solutions, we first need to understand which topology the network has, and what specific strengths and weaknesses that conveys. Whether we are talking about the networks of the banking system, mobile phones, international crime or the Internet, the science of complex networks evidently has a lot to teach us.

Further Reading

A.-L. Barabási, Linked: *The New Science of Networks*. Perseus, Cambridge, Ma., 2002.

D. J. Watts, Six Degrees: *The Science of a Connected Age*. W. W/ Norton & Co., New York, 2004.

M. Buchanan, Nexus: *Small Worlds and the New Science of Networks*. W. W. Norton & Co, New York, 2002.

M. Newman, Networks: *An Introduction*. Oxford University Press, 2010.

M. Newman, A.-L. Barabási & D. J. Watts, *The Structure and Dynamics of Networks*. Princeton University Press, Princeton, 2006.

A. Barrat, M. Bathélemy & A. Vespignani, *Dynamical Processes on Complex Networks*. Cambridge University Press, Campbridge, 2010.

A.-L. Barabási & E. Bonabeau, 'Scale-free networks', *Sci. Am.* **288**, 60–69 (2003).

A.-L. Barabási, 'The architecture of complexity', *IEEE Control Syst. Mag.* **27**(4), 33–42 (2007).

J. Gao, S. V. Buldyrev, H. E. Stanley & S. Havlin, 'Networks formed from interdependent networks', *Nat. Phys.* **8**, 40–48 (2012).

A. Vespignani, 'The fragility of interdependency', *Nature* **464**, 984–985 (2010).

N. Gulbahce & S. Lehmann, 'The art of community detection', *BioEssays* **30**, 934–938 (2008).

Spreading It Around: Mobility, Disease and Epidemics

The Black Death, generally thought to be bubonic plague, spread into Europe from the Caspian Sea, brought by Mongols attacking the city of Kaffa in 1346. Carried by fleas on ship rats, it was ferried to ports throughout Europe. Over the next seven years it killed about a third of the population.

The transmission of an infectious disease in the Middle Ages is fairly predictable. While ships could transport the plague over long distances, its spread across the European continent from the Mediterranean ports resembles the steady advance of an ink blot. Human mobility was then very low: most people barely ventured a few miles beyond their hometown, and so infection advanced more or less village by village.

It's different today. Individuals cross the globe in less than a day, while road, rail and water transport also allow rapid, long-distance movements. Increased human mobility recently has become a focus of attention because of fears about the spread of particularly virulent forms of influenza, such as H5N1 (bird flu) and H1N1 (swine flu). Yet despite recognition of how much more mobile we are than in the fourteenth century, many models of epidemics still assume that diseases spread in smoothly advancing fronts. Largely this is due to a sheer lack of information about what patterns of human movement really look like.

Once these patterns are taken into account, the complexity of epidemiology is greatly increased: what seems at first like a purely medical question becomes linked to quite different areas of social science, such as the nature of transportation networks and their patterns of usage. At a local scale, modeling of epidemics might need to take account of the kinds of human movement models described in Chapter 2, which could determine for example how many people we encounter in our daily routines. Human movements, the nature and patterns of our social (and indeed sexual) intercourse, and variations in susceptibility to disease must also acknowledge the influence of socio-economic demographics: affluence/poverty and culture

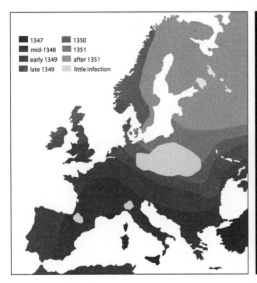 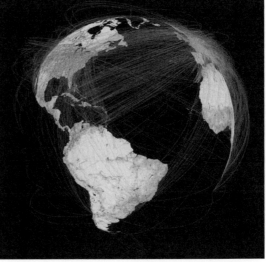

When human mobility was low, infectious diseases such as the Black Death spread slowly and predictably, like an ink blot (left). Today, humans criss-cross the globe in an instant (right), with potentially dire consequences for disease epidemics. (Credit (right): from B. Balcan et al., *Proc. Natl. Acad. Sci. USA* **106**, 21484-21489 (2009).)

affect all these things. In short, understanding the spread of disease today demands a truly cross-disciplinary perspective that, among other things, recognizes the complex interplay of many networks and modes of interaction.

The vital and urgent importance of that objective is in no doubt. AIDS is now the third biggest cause of premature death in the world, and kills two million people a year in sub-Saharan Africa alone. Around one in 20 people in that region aged between 15 and 49 are HIV-positive, reaching a level of almost one in seven in southern Africa. The disease is partly responsible for a life expectancy of around 50 and for preventing economic growth in the continent. Meanwhile, epidemiologists generally agree that a new flu pandemic, comparable to those in 1957–8 and 1968–9 that killed millions worldwide, is inevitable and yet impossible to predict. Although combating such major health problems is partly a question of developing new drugs and understanding the biology of the pathogens, more than ever before it relies on understanding how the diseases are transmitted and spread through patterns of human movement and behaviour. Only then, for example, can effective vaccination and quarantine strategies be devised.

Getting About

In 2006 researchers in Germany described an ingenious way of assessing how people move throughout the US by using banknotes as proxies. An amateur online database called wheresgeorge.com (referring to the picture of George Washington on a $ 1 bill) records the whereabouts of specific dollar banknotes, which are logged according to serial number by volunteer users. These entries obviously provide a highly incomplete record of the trajectory of any particular bill, but they offer an approximate indication of where and when people move from place to place: a bill usually changes hands when it is physically carried between transactions.

If these movements were like the diffusion of ink particles in water, a bill would travel a distance from its original location that is proportional to the square root of the time elapsed. But that's not what the researchers found. While many bills travel only short distances between successive reports on wheresgeorge.com, some jump hundreds of kilometres in that interval, presumably because someone has traveled far afield with them before using them to pay for goods or services. On average, for relatively short time spans the probability that a bill has traveled a distance r within that time is proportional to 1/r raised to the power of about 1.6: a power law (see page 25). This sort of spreading behaviour is characteristic of entities traveling not by diffusion but by a so-called Lévy flight, in which the short random hops of diffusion are interrupted every so often by 'scale-free' jumps, which can be of any size.

Monitoring mass human movements directly, rather than by proxies of uncertain representativeness, is becoming ever easier because we carry with us wireless detectors of our location, such as mobile phones. Telecommunications companies collect data about where individual calls were made, in part to monitor the completeness of their network coverage but also because the location of the user's nearest mobile-network tower can affect billing. Some companies make this anonymous data available for research purposes. A study of 100,000 people tracked via their mobile phones over six months showed that individuals' movements are generally highly predictable, and that we regularly return to a few locations and spend most of our time there. What's more, we typically make phone calls in a 'bursty' fashion – not spaced out evenly over time, but in groups interspersed with periods of inactivity. This seems to be a very common pattern in

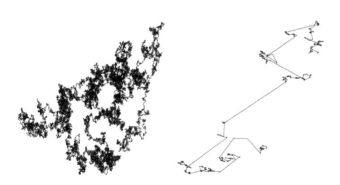

While a particle moving diffusively follows a 'random walk' – a series of small steps in random directions (left), human movements seem more like so-called Lévy flights, where the small steps are occasionally interrupted by large ones (right). (Credit: Courtesy of Diederick Wiersma, Università degli Studi di Firenze.)

human behaviour, seen also for example in the way we respond to letters or emails.

Such phone-tracking data show that, while people do seem to follow Lévy-flight trajectories, there is a lot of person-to-person variation in their precise characteristics: we each have our own 'signature' variation on the same basic pattern. People who are closely linked in a social network – who call each other frequently – also show similar patterns of mobility: they tend to frequent the same places. And the chance of forming a new link between two nodes in a social network is considerably greater if those nodes are already close together, separated by only a few steps – or equivalently, if they display similar mobility trajectories. Thus, studying mobile-phone data can not only tell us about people's physical movements in space but can also serve as a proxy for uncovering social network structures. This could be useful, not least for epidemiological modeling, but it also has a surprising and sobering implication from the perspective of privacy: purely by virtue of how we move around, we unwittingly encode and broadcast information about who we are and who we know.

Going Viral

The transmission of some infectious diseases, such as AIDS and other sexually transmitted diseases, requires more intimate contact than merely transient proximity. The dynamics of infection are therefore determined not so much by how people move around as by what their networks of social, physical and sexual contact look like.

Biologists have studied the transmission of disease through populations for over a century, but it is only recently that they have started to realize how important a role the topology of these network plays – in other words, that it matters what shape the branching has. One of the most widely used of the traditional models for studying epidemics, called the susceptible-infected-susceptible (SIS) model, supposes that individuals come in two classes: healthy and infected. If a healthy person encounters an infected one, they have a chance of becoming infected. Meanwhile, infected individuals can recover and become healthy again. In this model, the disease spreads at a rate that depends on the relative probabilities of infection and of recovery. If this spreading rate exceeds some threshold value, the disease becomes an epidemic, sweeping through the entire population and persistently infecting some constant proportion of them. If the rate is smaller than this threshold, the disease dies out. An epidemic can be avoided by keeping the spreading rate below this threshold level, for example by vaccinations that reduce the infection probability.

But if the encounters between people are described by the kind of 'scale-free' network that characterizes many real-world social networks (see Chapter 5), the outcome of the SIS model is rather different. There is no longer an epidemic threshold: all diseases can pervade the network no matter how slow-spreading they are. (In fact, despite their 'small world' interconnectedness, spreading on scale-free networks can be surprisingly slow, because it doesn't necessarily take advantage of the short cuts.) That seems to be the case for computer viruses, which are transmitted by messages passing through the scale-free email network: they linger at a low level for long times by persistently infecting a very small fraction of computers.

Sexual contacts define the network on which the AIDS virus HIV spreads, and to develop effective strategies for attacking the epidemic we need to know its topology. This is still under debate. A study of 3,000 Swedes suggested that the distribution in the number of sexual partners over a twelve-month period follows a power law: the signature of a scale-free network. But other studies have questioned this conclusion. All the same, it seems likely that networks of sexual contact share with scale-free networks the property of having a few very highly connected hubs – highly sexually active individuals – that can potentially infect a great many others. Several of the earliest cases of AIDS in Europe were tracked to a Norwegian sailor who contracted it in West Africa in the 1960s and then passed it on during a promiscuous career as a truck driver in Germany in the 1970s.

These 'hubs' are the key to an effective strategy for combating infectious diseases that spread on such networks. By targeting immunization or preventative treatments at such key individuals, the chances of an epidemic developing can be greatly reduced without the need for expensive or impractical mass prophylactic programs. Even if it can be hard to locate the most highly connected individuals accurately, their influence is so great that a relatively inefficient 'targeted' campaign can raise the epidemic threshold significantly.

However, a high degree of connectivity in the contact network may not be the only, or even the main, feature that determines the influence of individuals in the spreading of infectious diseases. For example, if a 'hub' is situated only at the periphery of the network, such that it has many local connections but few that link it into the main body of the web, it won't play such a big role. Less well-connected nodes at the core of the network may be more significant. Moreover, if spreading starts from several locations at once, the extent of spreading depends on how far apart they are. So targeted immunization strategies may need to consider quite carefully the overall shape of the network and the locations of individuals within it.

epidemic modeling is to adapt itself to the public-health challenges of the coming century.

Contagious Behaviour

Ultimately, even a more sophisticated view of the patterns of human mobility and contact aren't enough for accurately anticipating how epidemics and pandemics may spread. For one thing, faced with the threat of serious or fatal disease, people may change their behaviour. This was clearly the case for AIDS, where awareness of the dangers transformed the nature of sexual interactions in many communities. It was also evident in changing travel patterns to Asia during outbreaks of bird and swine flu. Moreover, cultural norms can influence the ways a disease might be transmitted and the responses to preventative strategies. For example, gender differences in attitudes to condom use, religious beliefs, and politically motivated 'HIV denial' have all strongly impacted the effectiveness of AIDS prevention in Africa.

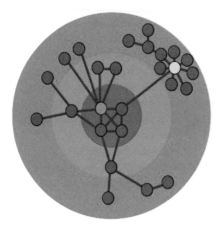

The importance of hubs depends on their position in the whole network. Here the blue and yellow hubs both have a high degree of connectivity: eight links. But the blue one, lying in the core of the network (red region), has a much greater impact on spreading through the network than the yellow one at the edge (blue region).

Findings like this have major implications for strategies to combat contagious disease. They illustrate that an understanding of social complexity is urgently needed if

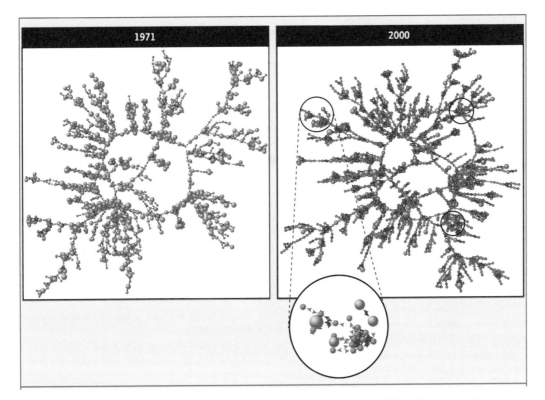

Part of the social network for a medical study conducted between 1971 (left) and 2000 (right). Each circle represents a person: yellow borders denote women, and red borders men. The interior colour and size indicate daily cigarette consumption – yellow for smokers (with a size proportional to intake), green for non-smokers. Orange links denote a friendship or marital tie, purple a familial tie. By 2000, most smokers appear at the peripheries of the network, in relatively small subgroups. The black circles in the 2000 network show some densely connected clusters of predominantly non-smokers, with a few smokers on the periphery. (Credit: from N. A. Christakis & J. H. Fowler, *New Engl. J. Med.* **358**, 2249–2258 (2008).)

This suggests that understanding and predicting the spread of diseases are multi-dimensional problems that need to integrate broader social dynamics, opinion formation, the readiness and resources of medical services, and other factors not generally incorporated into today's epidemic models. There are already the beginnings of such efforts, for example in the Global-Scale Agent Model developed at the Brookings Institution in Washington DC. An agent-based social-behaviour model devised at the US National Institutes of Health was also used to shape policy towards bird flu, and to devise containment strategies in the event of a terrorist-induced outbreak of smallpox.

One important aspect of such models will be the kind of collective decision-making discussed in Chapter 3. 'Fear epidemics' such as that witnessed for the MMR vaccine in the UK, which was spuriously linked to autism, and that which still threatens to affect flu vaccination programs in the US, could undermine otherwise well-planned preventative regimes. Peer pressure can inhibit rational responses or objective assessments, leading to runaway feedbacks analogous to the crowd-following or herding behaviour that may cause economic crashes (see Chapter 7).

This element of social interaction can even give a health issue such as obesity – which is becoming increasingly prevalent in many affluent nations – the characteristics of an infectious disease. Not only do obese people form clusters in the social network, but these clusters function more as cause than as consequence: people are more likely to become obese if their friends or siblings do, as opposed to people who are already obese preferentially forming social ties. One study revealed that a person's chance of becoming obese increases by 57 % if one of their friends does. The reasons for this 'contagion' of obesity are not yet fully clear – it could for example be simple mimicry of eating habits, or increased tolerance of weight gain, or even a physiological effect of imitation-induced changes in the brain's ability to impose limits on food intake. In any event, regarding this kind of behaviour-governed health problem as if it were a contagion could make a big difference to the effectiveness of preventative campaigns. As social scientists Nicholas Christakis and James Fowler have put it, "people are connected, so their health is connected."

Conversely, healthy behaviour can spread through social networks too. Non-smoking appears to have done so. The prevalence of smoking in the US has declined substantially in the past four decades (from 45 % of adults and young people to 21 %). A network analysis of over 12,000 people showed that smokers tend to cluster together and that the size of these clusters has remained more or less unchanged during that time. So the decline seems to be caused by cessation of smoking in entire social clusters. This study found that a smoker has a 67 % smaller chance of smoking if their spouse quits, 25 % smaller if a sibling quits, and 36 % if a friend quits. This emulation behaviour was stronger for people with higher levels of education. And the remaining clusters of smokers were located increasingly on the periphery of the social network, showing which individuals and groups need to be most targeted in anti-smoking programmes. It's worth bearing in mind too that, not only does the increasing complexity and interconnectedness of our world suggest and demand new ways of thinking about problems such as this, but it also promises to assist the implementation of solutions: information and communication technologies make it increasingly feasible to target health information at specific individuals.

Further Reading

J. Taubenberger, D. Morens & A. Fauci 'The next influenza pandemic: can it be predicted?' *J. Am. Med. Assoc.* **297**, 2025–2027 (2007).

J. M. Epstein, 'Modelling to contain pandemics', *Nature* **460**, 687 (2009).

D. Balcan, V. Colizza, B. Gonçalves, H. Hu, J. J. Ramaso & A. Vespignani, 'Multiscale mobility networks and the spatial spreading of infectious diseases', *Proc. Natl Acad. Sci. USA* **106**, 21484-21489 (2009).

D. Brockmann, L. Hufnagel & T. Geisel, 'The scaling laws of human travel', *Nature* **439**, 462–465 (2006).

L. Hufnagel, D. Brockmann & T. Geisel, 'Forecast and control of epidemics in a globalized world', *Proc. Natl Acad. Sci. USA* **101**, 15124–15129 (2004).

M. González, C. Hidalgo & A.-L. Barabási, 'Understanding individual human mobility patterns, *Nature* **453**, 779–782 (2008).

D. Wang, D. Pedreschi, C. Song, F. Giannotti & A.-L. Barabási, 'Human mobility, social ties, and link prediction", *ACM SIGKDD International Conference on Knowledge Discovery and Data Mining (KDD)* (2011).

F. Simini, M. C. González, A. Maritan & A.-L. Barabási, '*A universal model for mobility and migration patterns*', arxiv preprint 1111.0586.

R. Pastor-Satorras & A. Vespignani, 'Epidemic spreading in scale-free networks', *Phys. Rev. Lett.* **86**, 3200–3203 (2001).

R. Pastor-Satorras & A. Vespignani, 'Immunization of complex networks', *Phys. Rev. E* **65**, 036104 (2002).

A. Vespignani, 'Predicting the behavior of techno-social systems', *Science* **325**, 425–428 (2009).

N. A. Christakis & J. H. Fowler, 'The spread of obesity in a large social network over 32 years', *New. Engl. J. Med.* **357**, 370–379 (2007).

N. A. Christakis & J. H. Fowler, 'The collective dynamics of smoking in a large social network', *New Engl. J. Med.* **358**, 2249–2258 (2008).

M. Kitsak, L. K. Gallos, S. Havlin, F. Liljeros, L. Muchnik, H. E. Stanley & H. A. Makse, 'Identification of influential spreaders in complex networks', *Nat. Phys.* **6**, 888–893 (2010).

After the Crash: Economic and Financial Systems

"Economic predictions are notoriously unreliable", wrote the Nobel laureate economist Amartya Sen in 1986. "It is, in fact, tempting to see the economist as the trapeze-performer who tends to miss the cross-bar, or as the jockey who keeps falling off his horse." In October of the following year the stock market crashed on 'Black Monday' – and like all previous crashes, it came as a surprise to almost everyone.

The poor track record of economists in forecasting major shocks like this is now routinely cited as vindication of Thomas Carlyle's famous (and usually misunderstood) characterization of economics as "the dismal science". But this may be unfair. If ever there was a subject demonstrating how inappropriate it is to label the social sciences 'soft', it is economics. Unlike the 'hard' science of physics, whatever laws there might be that govern economic behaviour, they seem sure to be context-dependent, partial and inconstant over time.

All the same, the catastrophic consequences of the global financial crisis that began in 2008 have strengthened a growing conviction that we can't carry on this way – and that something is missing from the conventional economic models that prevents them from describing, let alone predicting, such serious deviations from normality. Perhaps we're wrong even to consider crashes (and bubbles) to be distinct from economic normality in the first place. Given that they have always existed, might they not instead be intrinsic features of the way markets work?

It is no mystery that conventional economic theories – such as those widely employed by central banks and economic institutions that inform government policies – fail to anticipate market crashes. For these theories are systematically constructed to *exclude* the very existence of such events. They insist that the economy is prevented from operating in a stable, regular fashion by disturbances that originate outside the economic system itself, and which therefore cannot possibly be taken into account by the models.

There has in the past several decades been a growing readiness to modify or even set aside these conventional concepts in favour of a recognition that the global economy is an immensely complex system, best studied and modelled by taking advantage of the insights gleaned from other facets of the science of complexity, whether these be in ecology, behavioural biology or physics. This perspective is still a minority view. But it has already demonstrated its worth and its potential benefits, and has been endorsed by some leading figures in economics.

The 2008 crisis ought to mark a turning point. The credit crunch and the ensuing national debt crises have revealed more clearly than ever how many of the phenomena now familiar from other areas of social complexity science also operate in the economic system. Hierarchical networks of interdependency, cascading breakdowns, herding behaviour and collective opinion formation, feedbacks that create extreme sensitivities to small perturbations – all have been implicated in the latest, tumultuous crash and its continuing and alarming repercussions. Moreover, the crisis shows the true financial and social cost of ignoring these considerations. If massive investment in a science of economic complexity were to relieve the consequences of events such as the 2008 crisis by only a percent or so – let alone predicting and avoiding them – then the expenditure will have been justified many times over.

The Problem with Economics

Traditional economic theory makes several fundamental assumptions that seem now to be excessively simplistic. The first is to imagine that the economy is an equilibrium system. In other words, in the absence of confuting influences from 'outside', supply and demand would find a perfect balance everywhere so that all markets would clear: there would be no surpluses or supply shortfalls, and prices would be stable. This assumption stems from the origins of microeconomic theory as an analogue of theories of equilibrium physical systems such as gases, which have stable, unchanging states. The physical sciences have long since moved on to describe non-equilibrium processes such as the weather system (in which chaotic behaviour was identified in the 1960s), but economics has not. The implications are huge. The 'equilibrium paradigm' explains why the so-called dynamic stochastic general equilibrium models prevalent in economic forecasting ignore the potential for major fluctuations such as slumps and crashes. It also motivated the disastrous suggestions by many politicians before the 2008 crisis that such crises had been banished for good.

Conventional models make additional over-simplifications. At their most basic, they state that all agents in the economy are identical, and that all have access to total information about the economy, on the basis of which they make the rational choices that will optimize their 'utility': maximizing revenue or profits, say, or finding an ideal work/leisure balance. These assumptions have been relaxed in various ways by more sophisticated models, which for example recognize inequalities in the information that agents can access, or bounds on their ability to reach a rational optimum decision. However, the recent emergence of experimental behavioural economics – which examines how people really behave in making transactions – has revealed the gulf that still exists between the behaviour that models assume and that exhibited by real people.

Also notably lacking in these models is an acknowledgement of feedbacks and interdependencies of behaviour. Agents only interact with one another via the indirect mechanism of how their decisions affect prices. The fact that fluctuations, bubbles and even crashes can be driven by herd-like copying – irrationally inflated asset prices or panic selling, for example – is widely remarked in the financial press but rarely admitted into models. As the economics Nobel laureate Joseph Stiglitz put it in 2008, "Many of the problems our economy faces are the result of the use of misguided models. Unfortunately, too many [economic policy-makers] took the overly simplistic models [used in] courses in the principles of economics (which typically assume perfect information) and assumed they could use them as a basis for economic policy."

These shortcomings are precisely the kinds of things that models based on complex systems are well placed to handle. One of the most promising alternatives to the pen-and-paper equations that are typically used to describe an equilibrium economy ruffled by random noise are agent-based models, in which the assumptions are not top-down conditions such as economic equilibrium or perfect information but rather, the rules of interaction and trading of individual agents. There are far fewer top-down (and often quasi-ideological) assumptions built into these models about the gross nature of the economic markets. Rather, one simply observes what aggregate behaviour the 'microscopic' rules produce. As economist Alan Kirman has put it, agent-based models "provide an account of macro phenomena which are caused by interaction at the micro level but are no longer a blown-up version of that activity."

Such economic models typically ascribe certain decision-making rules to each agent – which need not be, and in general are not, identical from one agent to the next. Given a set of prices, for example, the agents might each apply different rules of thumb in deciding which transactions to make. These rules may take account of what others are doing. They could be probabilistic rather than deterministic: given a certain set of circumstances, there could be a 70 % probability that an agent will take action A, and a 30 % chance of action B. As in real life, some agents could make their decisions on the basis of calculations, others by looking at past trends, and so forth.

Crucially, the agents may respond to one another, enabling the copycat behaviour that leads to herding and other collective actions. They may learn from experience, or switch their strategies according to the majority opinion. They can aggregate into institutional structures such as banks and firms. These things are very hard, sometimes impossible, to build into conventional models.

With so many possibilities for how to describe and quantify agent behaviour, can anything general be said about the outcomes? Yes, it can. In agent-based models, the economy that emerges tends to be out of equilibrium: it never settles down into a steady state. The fluctuations in pricing are then not imposed by outside disturbances (exogenous) but are an intrinsic aspect of the system (endogenous), resulting from the intricate web of interactions and feedbacks between agents, just like the variability in the weather. And in the same way, this variability – or what economists call volatility – might be greater in some circumstances, or some sectors, than in others.

One of the most significant aspects of fluctuations in agent-based models is that they tend to happen on all scales. Day-to-day variations in prices are usually small, but occasionally they can reach big peaks or lows. These are generally not related to changes in the intrinsic value of those commodities (although of course in reality asset prices do feel the effects of exogenous influences, for example if a technology company benefits from a scientific breakthrough), but are the collective result of many individual decisions. These fluctuation statistics contrast with those usually imposed on conventional models by injecting random 'white noise' into the equations: they look more like the 'scale-free' statistics seen in the real economy, and moreover tend to be manifested in bursts just like real-world 'volatility clustering'. In short, they look realistic.

This is not a mere technicality. A correct description of fluctuations is, for example, an essential ingredient in theories for pricing and risk assessment of derivatives. The conventional assumption that these are like white noise can lead to dangerously inaccurate results.

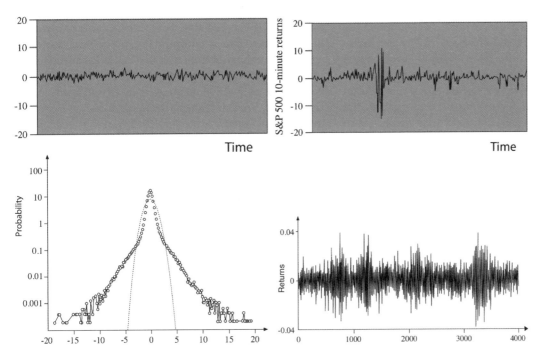

In traditional models, economic fluctuations are often assumed to be like white noise (top left). But in fact they tend to have bigger spikes than that (top right). A plot of the statistical distribution (lower left, here for fluctuations of the Standard & Poor's 500 market index) shows how different the real variability is from white noise, which would give the dashed curve. Agent-based economic models typically produce more realistic fluctuations with more large spikes: an example is shown in the lower right.

Really large fluctuations like those seen in bubbles and crashes appear automatically in these models as an aspect of the endogenous complex behaviour. In traditional models that assume white-noise variability, these very large events are so rare that to all intents and purposes they never happen. In contrast, agent-based models suggest that such events are a recurrent, albeit uncommon, feature of how the economy works. Crises are 'natural' to them. It was a failure to acknowledge the true likelihood of 'rare' big fluctuations that ultimately lay behind the collapse of the British bank Northern Rock in the early stages of the credit crisis.

Agent-based approaches that venture beyond the classical paradigm of equilibrium economics are not – or should not be considered – in competition with that older picture. They are a generalization of it, an extension of the old methods. Given certain assumptions and conditions, the predictions of the out-of-equilibrium models can reduce to those of equilibrium models. This is as it should be, for some markets, under some conditions, do seem to be fairly well described by the traditional picture. But they are special cases, and not representative ones. Just as physicists and chemists have moved beyond a consideration of equilibrium phenomena to look at complex process that happen out of equilibrium, so it makes sense for social scientists, and especially for economists, to do so.

High Expectations

All economic decisions are made on the basis of expectations. A venture capitalist's decision to invest in a company depends on his or her expectations of how the company will fare in the future. Stocks are bought or sold on the basis of expectations about whether their value will rise or fall.

But there's a fundamental problem. These decisions don't just depend on the expectations, but affect whether the expectations are accurate or not. If no one invests in a company making a particular product, there will never be a market for it in the future. Prices of stocks tomorrow depend on how they are traded today. So these decisions are not only self-referential but require us to guess what other agents will decide.

Conventional economics cuts this Gordian knot with the assumption of rational expectations: there is a single best choice that can be determined on the basis of all the information available *now*, and which, if it is adopted by everyone, will lead to an outcome that validates the expectations which led to it. In short, this approach assumes that everyone knows the 'right' answer, that everyone will choose that answer, and that they will continue to do so indefinitely.

Unfortunately, not only is the probability of this happening as unlikely as it sounds, but there are some situations in which it is guaranteed to fail. One such is the so-called El Farol problem, in which agents have to decide whether or not to go to a popular bar (the El Farol, named after a bar in Santa Fe). If they assume that everyone else will go, then they will choose not to, so as to avoid an uncomfortably overcrowded evening. If they assume that others will stay away because of that very prospect, then they will go. But if there is a 'rational expectations' algorithm that tells everyone what to do, and everyone uses it, then everyone will make the wrong choice all the time, whether to go or stay away.

This situation has been explored by agent-based modelling. Each agent adopts a decision-making strategy, which may be different from others' and might incorporate past experience. What happens is that the attendance at the El Farol fluctuates continually, but converges to an average value in which some constant proportion of the agent population attends. It finds its own equilibrium by self-organization – but it is an equilibrium constantly prone to fluctuations, including occasional episodes of serious over-crowding or relative emptiness. This situation mimics a common one in economics in which agents seek to be in the minority – often a desirable position, for example being a buyer in a seller's market.

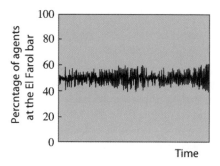

An agent-based model of the 'El Farol problem' (see text) converges on a constant average attendance but with constant and sometimes large fluctuations.

The Bigger Picture

While 'toy' models like the El Farol problem have established that an agent-based complex-systems approach can address economic questions that fox conventional theory, there are now ambitions that this approach should say something more specific about the markets and financial institutions of the real world. Unlike the microeconomic models still regularly used by institutions such as the US Federal Reserve, agent-based models can include banks, institutions and companies, modelling their interactions, their formation and growth, and their failure. They can give insights into *why* firms fail, and can offer explanations for the observed statistical distributions of firm size. Because they can accommodate interactions between agents, they may demonstrate the cascades and herding effects that give rise to economic bubbles and panics. In view of the critical role played in the current financial crisis by chains of interaction in bank lending and borrowing – the difficulty of tracing debt through this network, and its consequent vulnerability to breakdowns of trust – this ability to capture the structures of real markets could be crucial for better economic modelling and prediction.

Researchers are now discussing the feasibility of constructing an agent-based model capable of simulating a nation's or indeed the world's economy. That would be an immense undertaking: a whole-economy model would need to draw on the knowledge of economic experts in finance, labour markets, supply chains, marketing and retail, among others, as well as areas such as psychology and law, for example to elaborate the ground-rules that underlie human decision-making. Following the interactions of millions, or perhaps billions, of agents, companies and institutions with diverse agendas and decision-making rules would also demand huge computer power and would need vast inputs of data: patterns of spending and financial transactions, say, and detailed maps of the networks along which loans and risks are channeled.

The scale of the task would probably mean we'd need not one single model but many, whose collective forecasts could be pooled to map out possible futures. Such efforts have already begun. Between 2006 and 2009 a European team created a model called Eurace, the largest agent-based model of the economy developed so far. It simulated a fictitious economy with several million agents, including markets for labour, goods, credit and finance. Firms within the model were characterized as collections of 'worker' agents, and the model had an explicit spatial structure: firms and workers were located somewhere in real space, linked via social and business networks.

The Eurace team aim to use the model as a testing ground for aspects of European economic policy. For example, they were able to probe one of the most pressing questions today: how best to deal with massive governmental debts like those of Greece and Italy. Is the answer fiscal tightening, reducing the debt with high taxes or low public spending, or quantitative easing, keeping taxes low and plugging the debt by selling government bonds? Eurace's simulations suggest that in the long run economic growth is boosted and unemployment reduced more by the second approach – as long as firms are financially robust. More work is needed to prove that these and other preliminary results from Eurace scale up to the size of the actual national or international economy.

Agent-based models are not a panacea. There's still no general prescription for how to construct one that offers realistic, relevant and reproducible results, particularly in terms of what behavioural rules should guide the agents (how much psychological complexity they should include, for instance). It remains to be seen whether they can offer reliable policy advice on questions that traditional economic theories fail to tackle. Yet there has never been a clearer case for bringing the lessons of complexity science to bear on the behaviour of economic and financial markets. We are still living, perilously, with the failures of the traditional model: with an inability to predict the current crisis or to offer any kind of consensual and effective means of escaping it. It is now evident that events like these are not 'imperfections' that ruffle an equilibrium economy, but an intrinsic and deeply hazardous feature of the existing capitalist system, which can blight the economy for a decade or more. The existence of these fragilities can now be seen to stem from the influence of one economic actor on another, and on the structure of that network of interactions and its hidden vulnerabilities. Whether or not a complex-systems approach will ultimately succeed in taming the economy's worst convulsions, it would be nothing short of reckless not to try it.

Further Reading

W. B. Arthur, 'Out-of-equilibrium economics and agent-based modeling', in *Handbook of Computational Economics, Vol. 2: Agent-Based Computational Economics*, eds K. Judd and L. Tesfatsion. Elsevier/North-Holland, Amsterdam, 2006. [http://www.santafe.edu/~wbarthur/documents/OutofEquilPaper-SFI.pdf]

R. N. Mantegna & H. E. Stanley, *An Introduction to Econophysics*. Cambridge University Press, 1999.

A. Kirman, Complex Economics: *Individual and Collective Rationality*. Routledge, London, 2011.

D. Sornette, Why *Stock Markets Crash*. Princeton University Press, 2004.

P. L. Borrill & L. S. Tesfatsion, 'Agent-based modeling: the right mathematics for the social sciences?' *Working Paper* 10023, July 2010. Iowa State University, Ames.

B. LeBaron, 'Agent-based computational finance', in *Handbook of Computational Economics, Vol. 2: Agent-Based Computational Economics*, eds K. Judd and L. Tesfatsion. Elsevier/North-Holland, Amsterdam, 2006.

M. Gallegati & M. G. Richiardi, 'Agent-based models in economics and complexity', in *Complex Systems in Finance and Econometrics*. Springer, 2011.

B. LeBaron & L. Tesfatsion, 'Modeling macroeconomies as open-ended dynamic systems of interacting agents," *Am. Econ. Res. Papers & Proc.* **98**, 246–250 (2008).

J. D. Farmer & J. Geanakoplos, 'The virtues and vices of equilibrium and the future of financial economics', *Complexity* **14**, 11–38 (2009).

R. M. May, S. A. Levin & G. Sugihara, 'Ecology for bankers', *Nature* **451**, 893–895 (2008).

D. Farmer & D. Foley, 'The economy needs agent-based modelling', *Nature* **460**, 685–686 (2009).

N. Johnson & T. Lux, 'Ecology and economics', *Nature* **469**, 302–303 (2011).

T. Preis, J. J. Schneider & H. E. Stanley, 'Switching processes in financial markets', *Proc. Natl. Acad. Sci. USA* **108**, 7674–7678 (2011).

Love Thy Neighbour: How to Foster Cooperation

Society is a collaborative effort: it works to the extent that we can get along with our neighbours, agree on common goals, and accept shared responsibilities. Even opponents of state taxation generally recognize that there are benefits to the collective financing of public services – we don't want to build and maintain our own roads or hospitals. And because the growth of an economically disadvantaged 'underclass' can threaten the stability of society, a degree of redistribution of wealth benefits everyone. So while some of the most fundamental political divisions hinge on the question of where to draw the balance between collective responsibilities and individual liberties, all democratic societies acknowledge that their citizens have to some extent to find ways of cooperating with one another. In particular, liberal philosophers since the seventeenth century have concurred that civil peace and order come at the expense of individual restraint, including at the very least the renunciation of attempts to harm others.

Yet that seems to conflict with the supposed Darwinian imperative of competition, in which every individual is out for themselves. Long before Darwin's theory, some philosophers insisted that the only way to avoid the rapacious state of affairs that followed from humankind's greed and desire for power over others was to impose the restraining authority of the state. Others had more faith in humanity: they felt that God had made people inherently good and rational, and that this for the most part guarantees that our relations with our neighbours are civilized.

But appeals to the privileged 'rationality' of humankind are undermined by the fact that cooperation and altruism are seen also in the animal kingdom – for example, in pack hunting, grooming activity and the collective activities of social insects. Some of this behaviour was given a Darwinian explanation as it became understood that altruism among kinship groups can help to propagate genes at the expense of individuals, so that a genetic predisposition to cooperate with closely related individuals can foster survival at the genetic level. But not all altruism relies on kin relations.

Explaining how cooperation arises and is sustained in populations of self-interested agents has become a topic of intense interest among both evolutionary biologists and social scientists. The good understanding of this issue that has now developed is founded on the recognition that sociality arises from choices made by interacting agents who may come to realize the long-term benefits of cooperation with their neighbours. In other words, the explanation depends on considering the population as a complex system involving interactions and feedback.

The implications of these studies for social science are immense. Cooperation – its creation and its vulnerabilities – is arguably the defining feature of civilization. It is essential, for example, for the mutually beneficial exchange of goods, the payment of taxes, teamwork, management of common resources, collusion among firms, the reduction of socioeconomic inequality, participation in collective actions such as demonstrations, and adherence to socially beneficial norms. Many, perhaps even most, of the serious challenges and crises considered in this book – from crime and riots to the illiquidity of economic crashes, wars, and environmental degradation and global warming – are at root breakdowns of cooperation. All this makes it imperative for a science of social complexity to have a firm general understanding of how cooperation can arise and what factors can stabilize or undermine it.

In human society, cooperation is influenced and complicated by many other individual and collective traits, such as learning, reputation, incentives, the formation of institutions and laws, the capacity for punishment, and the interplay with conflict between rival groups. In some ways this means that the more we study cooperation, the more complicated it seems. On the other hand, many of these factors can now be incorporated into models and theories, and they are enabling the 'science of cooperation' to move ever further from idealized and perhaps over-simplified models towards descriptions that come close to capturing the complexity and subtlety of real-world behaviour.

Resolving the Dilemma

In one sense it doesn't seem hard to understand why, even in a starkly Darwinian world, we should be predisposed to cooperate. For there are many situations in which we can accomplish far more through group effort than through uncoordinated individual action. Lone hunters in the Pleistocene could not hope to kill large, savage beasts, and even a shared spoil is better than none. The fundamental difficulty is not so much finding a motivation for cooperative behaviour, but explaining why it does not constantly succumb to selfishness. Indeed, often it does. The 'tragedy of the commons' refers to the medieval tradition of grazing livestock on shared common ground, where individuals sometimes figured that no one would notice if they over-grazed their own herds As this tendency grew, eventually the land was left barren for all. The same is happening to fish stocks today.

This is the basic problem for societies based on the principle of unenforced cooperation: they are ripe for exploitation by those who put self before community. In this sense, the policing role of the state might be considered not so much a question of forcing everyone to cooperate against their instincts, but of protecting against 'defectors' and free-riders. This dilemma has been recognized for centuries, but it was framed in a formal, scientific way in the 1950s within the context of the nascent discipline of game theory, which seeks a description of behaviour in which individuals anticipate how others will act towards them. The classic model system for investigating cooperation and defection is a 'game' called the Prisoner's Dilemma, in which two players are presented with the choice of whether to cooperate or not (that is, to 'defect'), with the temptation that, while mutual cooperation is a good outcome for both players, unilateral defection has an even better payoff for the defector (see Box below). A rational analysis of the options indicates that, regardless of what the other player does, it is always better to defect. Yet this must then lead to both prisoners defecting, which is a worse outcome than if they both cooperate.

If the game is repeated many times, it should therefore eventually dawn on truly rational players that their long-term interests are best served by cooperating. Thus self-interest can promote apparent altruism. This outcome of cooperation as a learnt response to the depredation of mutual defection was evident in the spontaneous truces that arose between entrenched forces in World War I. Both sides would cease shelling, or would fire shells obviously off target to maintain a pretense of combat, by silent mutual agreement. Here the emergence of cooperation relies on both parties expecting to continue to interact in the future, and being sufficiently able to anticipate what those future encounters could be like.

The Prisoner's Dilemma

The basic temptation to free-load or 'defect' in a cooperative group was recognized by Jean-Jacques Rousseau in the eighteenth century, who imagined a team collaborating on a stag hunt. When a hare comes within reach of one of the men, he grabs it – but without his help, the stag escapes. The 'defector' enjoys stewed hare, but his fellows have nothing.

In the 1950s two researchers at the RAND Corporation in California devised a simple mathematical model of a two-agent interaction, which became known as the Prisoner's Dilemma, that incorporated this element of temptation. The two players of the game are imagined as prisoners suspected of committing a crime. If one will testify against the other, he will receive a more lenient sentence while the other receives the harshest penalty. But if both testify against each other, they will both be more harshly sentenced than if both do not – that is, than if they cooperate with one another.

Logic dictates that it is always best for a prisoner to denounce the other (to defect), for the outcome is then better whichever choice the other prisoner makes. If prisoner 1 testifies and prisoner 2 does not, prisoner 1 gets off the charge – the best of all outcomes for him, while prisoner 2 is the 'sucker' who gets the worst outcome. If prisoner 2 testifies, on the other hand, then it is still better for prisoner 1 to testify than not: he gets a lighter sentence in the former case than the latter. So if the agents play 'rationally', they always get the meagre payoff of mutual defection, which is only just better than the sucker's payoff, and not as good as the reward of mutual cooperation.

		Player 2	
		Cooperate	Defect
Player 1	Cooperate	3 / 3	5 / 0
	Defect	0 / 5	1 / 1

■ The choices in the Prisoner's Dilemma can be expressed in terms of this table of quantitative payoffs for each pair of decisions. The exact numbers don't matter, apart from tuning the degree of 'temptation' – the point is in the ordering of benefits or penalties for each decision.

But just as unconditional defection is not a wise strategy, neither is unconditional cooperation. That works fine against another unconditional cooperator, but if a cooperative player comes up against a habitual defector, he will be exploited ruthlessly.

To identify the best strategy – the one that leads to optimal benefits over repeated rounds of the game – early studies of the Prisoner's Dilemma staged 'tournaments' in which agents using different strategies were pitched against one another. These studies found that the most effective strategy was a very simple one: make the same choice of cooperation or defection as your opponent made in the last round. This was called the Tit-for-Tat (TfT) strategy. It can lead to sustained coordination: against an unconditional cooperator, TfT always cooperates and never exploits. But it also punishes defection with defection: it is "tough but fair".

Patterns of Niceness

Subsequent studies of the Prisoner's Dilemma and related 'cooperation games' have revealed an immense amount of complexity in this simple model. For one thing, TfT is not by any means the 'perfect player'. There is no unique best strategy in the game – that depends in part on how the opposition plays, an illustration that the notion of an optimal, rational behaviour in an interactive situation like this has no real meaning unless the context, and particularly the behaviour of other agents, is taken into account. TfT itself is vulnerable to errors – one defection (perhaps a random mistake or 'misunderstanding') prompts another in an unending cycle. That sort of relentless reprisal is all too familiar from human conflicts such as those in the Middle East and formerly in Northern Ireland, and clearly frustrates any convergence to long-term peace. So in a game where 'noise' leads to some randomness in players' choices, a more forgiving strategy can perform better.

The context-dependence of outcomes is brought out in studies in which the agents are arranged in a spatial configuration such as a grid. This places constraints on the interactions that the players have: for example, cooperators that can draw mutual strength against defectors in a well mixed population might not be able to do so if locked into a grid: spatial isolation militates against cooperation, a conclusion that might be relevant to the foreign policies of Israel. Conversely, defectors can steadily 'colonize' a population of cooperators, often in complex spatial patterns that develop because defectors 'repel' other defectors (they fare poorly when pitched against each other). Depending on the precise payoffs of cooperation and defection, these spatial games can generate constantly shifting patchworks of cooperation and defection, some of them orderly and some chaotic. Moreover, studies with real players show that cooperative behaviour can spread in cascades through a social network: one person's 'good' behaviour can induce the same in another even if the two never interact directly.

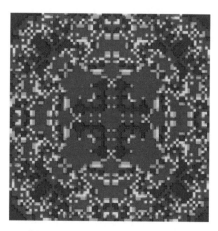

The spread of defectors (red) in a population of cooperators (blue) can exhibit complex spatial patterns. Here yellow and green sites denote agents that switched between cooperation and defection in the previous round. (Credit: from M. A. Nowak & R. M. May, *Int. J. Bifurcation & Chaos* **3**, 35 (1993).)

To investigate how behaviour evolves, many studies set up an evolutionary process in which agents can generate offspring that adopt the same strategy as their 'parents'. If those that are more successful – that achieve higher scores – produce more offspring, there is a kind of Darwinian selection for the best strategies: the 'fittest' multiply and the unfit die out. Equivalently, this could be posed as a process of learning from experience: agents are apt to copy strategies that prove successful for others. In other words, they change their behaviour in response to one another. These evolutionary Prisoner's Dilemma games can display complex time histories that are hard to predict – for example, exhibiting abrupt switches between predominantly cooperative and defecting populations, and concurrently, between the pre-eminence of different strategies. In one such study, widespread defection was overwhelmed suddenly by TfT, which itself then mellowed into a 'generous Tit for Tat' that forgave occasional defections: once this selfish society was tamed by tough reprisals, it could 'afford' to be more lenient. Such populations can eventually become even more forgiving, eventually to the extent that they become vulnerable to takeover by a few defectors that arise by chance – and so the cycle might continue, with 'nice' and 'nasty' periods following one another.

Cooperative interactions link us into networks much like those of friendship and collaboration. The two-way cooperation network between manufacturers and contractors turns out to have the same structural features, such as division into modules and statistics of link connections, as

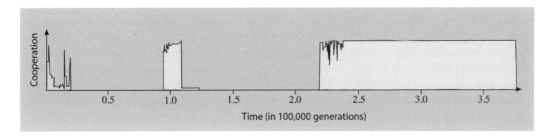

A time history of an evolutionary Prisoner's Dilemma game, showing the rise and fall of periods of cooperation (yellow).

that seen for insect-plant pollination co-dependency, suggesting that there may be simple, rather universal laws of how cooperation operates in very different settings both in human society and the natural world.

If a genetic predisposition towards altruism in humans did evolve in the way that these game-theoretical models imply, this does not necessarily imply a steady historical progression away from violence and towards cooperation. One agent-based model of group behaviour in early humans suggests that altruistic behaviour might initially have been parochial – concentrated within groups – and have been stimulated by inter-group conflict. If that's true, our genetic legacy might simultaneously predispose us both to cooperation and war. This somewhat pessimistic view does not have to provide a prescription for the future, however, for humans are highly susceptible to social learning: to evolving cultural modes of behaviour that can overcome our baser impulses.

Reputation Matters

One striking aspect of cooperative behaviour in the real world is that it need not take time to emerge via a series of repeated interactions between individuals. We tend to cooperate with strangers who we might not reasonably expect to see again. Evolutionary games are often underpinned by an assumption that tendencies towards cooperation can become genetically hardwired when cooperation is a successful strategy. This may well be so, but it is difficult here to separate what is innate from what is learnt or culturally acquired. It can also benefit individuals to cultivate a reputation for cooperation: to send out social signals that will enable them immediately to extract from new encounters the benefits that cooperation can offer. Reputation also helps to avoid a 'tragedy of the commons': it means that agents can contribute to the common good with a strong expectation that their cooperative behaviour will be reciprocated.

By the same token, being known for a tough but not necessarily automatic reprisal policy could be the best way to police defection. There can be a societal benefit in converting private transgressions into public information, for example via the procedures of civil trials, so that defectors are identified by bad reputations and their transgressions do not threaten general dissolution of the important social capital of trust.

The function of reputation as the midwife of cooperation has become apparent empirically in online markets such as eBay, where it was instituted through a lucky hunch. But while one can always construct plausible narratives to explain such effects, there is nothing obvious or inevitable about them. Only by modelling these social interactions as a complex system can one hope to develop a systematic understanding of how, why and when such measures will be effective.

Crime and Punishment

Political philosophers such as Hobbes and Locke considered that one of the functions of the state is to enact reprisals on defectors so that cooperators don't have to do it themselves. A police force relieves us from the need to meet aggressors with violence, for example via brutal vigilante groups. But recent studies in which human participants play 'games' involving cooperation and defection have shown that punishment seems to satisfy a more deep-seated need. Players typically exhibit a strong sense of justice that makes them willing to punish defectors even at a cost to themselves.

In these studies, punishment is often examined in so-called public-goods games, where players are asked to pool a resource – usually real money, which makes them take the game seriously. The more they contribute, the more they are rewarded. But freeloaders can benefit from the share-out of rewards without themselves contributing. Players are usually insistent on such behaviour being punished, even if in effect they have to pay for it. In other words, there is more involved in these decisions than a hard-headed 'maximization' analysis.

How severe does punishment need to be? One model

of a public-goods game suggests that fostering cooperation may depend on the strength of punishment in subtle, non-intuitive ways. Above a critical punishment threshold (the size of the fine imposed for defection, say), cooperators who punish can gain strength by sticking together, eventually crowding out both defectors and non-punishing cooperators (who can also be considered a kind of free-rider). But if punishment is carried out not by cooperators but by other defectors, too high a fine is counterproductive and reduces cooperation. Defectors who punish other defectors not only are found in behavioural experiments but are also familiar in reality: there are both 'hypocritical' punishing defectors (evangelists whose condemnation of sexual misdemeanours ignores their own, say) and 'sincere' ones, who deplore certain types of cheating while practising others.

From Games to Reality

Game theory and its implications for cooperative behaviour have already influenced political policies. But its early, simplistic variants offered no guarantee of sound advice. The original Prisoner's Dilemma itself, for example, seemed to imply that a defection strategy was the best course for Cold War nuclear-weapons policy: unilateral armament, and arguably even a first-strike agenda. Over time, the crippling cost and the catastrophic sensitivity to error made cooperation – arms limitations and disarmament – seem the better option.

It seems very likely that behaviour arising initially from rational 'best choices' will become over time reinforced by cultural norms that are divorced from their original raison d'être. And of course people's decisions about 'cooperation' and 'defection' are seldom made in context of single, transparent and quantitative payoffs, nor do they tend to involve black-and-white options for aligned or opposed behaviour. We are guided by a complicated mixture of moral and ideological preconceptions, conflicting impulses, social pressures (will we behave 'well' if no one will witness our trespasses?) and changing circumstances (would poverty make us more inclined to steal?).

Nonetheless, the strong resonances already apparent between the predictions and outcomes of cooperative games and real-world behaviour – for example, the benefits of solidarity for altruism, the need for deterrents to free-riders, and the collective evolution of norms of mutual help or selfishness – suggest that certain social traits might already be captured to some degree in these models, and that there is already a sound foundation on which to build a deeper understanding of how societies succeed or fail in developing a cohesive and collaborative community.

Further Reading

R. Axelrod, *The Evolution of Cooperation*. Basic Books, New York, 1984.

M. A. Nowak & R. Highfield, *SuperCooperators: Why We Need Each Other to Succeed*. Simon & Schuster, New York, 2011.

M. Milinski, D. Semmann & H.-J. Krambeck, 'Reputation helps solve the 'tragedy of the commons'', *Nature* **415**, 424–426 (2002).

S. Saavedra, F. Reed-Tsochas & B. Uzzi, 'A simple model of bipartite cooperation for ecological and organizational networks', *Nature* **457**, 463–466 (2009).

D. Helbing, A. Szolnoki, M. Perc & G. Szabó, 'Punish, nut not too hard: how costly punishment spreads in the spatial public goods game', *New J. Phys.* **12**, 083005 (2010).

D. Helbing, A. Szolnoki, M. Perc & G. Szabó, G., 'Evolutionary establishment of moral and double moral standards through spatial interactions', *PLoS Comput. Biol.* **6**, e1000758 (2010).

D. Helbing & A. Johansson, 'Cooperation, norms, and revolutions: a unified game-theoretical approach', *PLoS ONE* **5**, e12530 (2010).

D. Helbing & W. Yu, 'The outbreak of cooperation among success-driven individuals under noisy conditions', *Proc. Natl. Acad. Sci. USA* **106**, 3680–3685 (2009).

E. Fehr & S. Gächter, 'Altruistic punishment in humans', *Nature* **415**, 137–140 (2002).

A. Dreber, D. G. Rand, D. Fudenberg & M. A. Nowak, 'Winners don't punish', *Nature* **452**, 348–351 (2008).

S. Bowles & H. Gintis, 'Cooperation', in *The New Palgrave Dictionary of Economics*, eds L. Blume & S. Durlauf. Macmillan, London, 2008.

J.-K. Choi & S. Bowles, 'The coevolution of altruism and war', *Science* **318**, 636–640 (2007).

Living Cities: Urban Development as a Complex System

9

A United Nations report in 2007 announced that more than half the world's population now lives in cities. This shift in the balance between urban and rural dwelling is unprecedented in human history, and implies that for most of humankind the future is an urban one. Over the past several decades there has been massive migration of people from the countryside into cities. Partly as a result, there are now many mega-cities with populations of over 10 million, most of which are in developing countries in Asia, Africa and South America. Nearly all the population growth forecast for the next two decades will be based in such cities.

Given that cities have been growing for centuries, these demographic changes might seem like just more of the same. But they are not. They present many new and serious challenges, for which past experience will be a poor guide. For one thing, the pace of the changes is much greater than in the past: in China, for example, the exodus to cities has been greatly stimulated by the country's recent economic acceleration. Climate and environmental change are also predicted to force many rural dwellers to find a new urban livelihood – even though those cities often lie on coasts or floodplains that could actually be exposed to greater environmental risk. The aims of and constraints on urban development have also shifted: today there is a recognition that such changes need to be managed in a sustainable way, with a balance found between several social, economic and environmental factors. As one urban theorist put it recently, "the planner must reconcile at least three conflicting interests: to 'grow' the economy, distribute this growth fairly, and in the process not degrade the ecosystem."

Yet already the infrastructures for water supply, sanitation, transport, energy and health are inadequate to meet the needs of many cities in the developing world. At the same time, cities have been relatively neglected in poverty relief programs. According to Hans van Ginkel, former rector of the United Nations University in Tokyo, "Public officials and researchers have often underestimated, or even denied, the importance of cities... This partly explains why policy interventions rarely address the root causes of urban problems, and why, in some cases, the policies are misguided."

Besides all this, cities in the future will not necessarily resemble those today. Indeed, the whole concept of a city is likely to become hazy: they will not be well defined entities with clear boundaries, but rather, webs of urban development of varying size, density and function. "This new 'melded' landscape, characterized by the emergence of large populated regions interacting with their hinterlands and beyond, in ever-more complex and kaleidoscopic patterns, represents our urban future", says van Ginkel. "There is no escape from it."

Partly these developments tend to be framed as a technical challenge, requiring new means of ensuring clean water, affordable housing and so forth. But the challenge is also, and perhaps more significantly, a conceptual one: can we understand how cities evolve, and how they will do so in the future? An integrated view of urban growth is essential for managing these expected changes. It's not enough to consider, say, transportation networks in isolation from energy distribution or the patterns of wealth and age demographics. One of the messages emerging from studies of urban growth is the same that we have heard earlier in other fields of social complexity: it is no longer tenable to imagine planning and prediction as a matter of top-down control. The only effective way to manage cities will be to discover their intrinsic bottom-up principles of self-organization, and then to work with those so as to guide the process along desirable routes, rather than trying to impose some unreachable or unsustainable order and structure.

City as Organism

Cities have since the 1950s been viewed as complex systems whose organization and structure is the result of a hierarchy of lower-level components. But the traditional view was that these were 'controlled' systems, maintained in an equilibrium state by negative feedbacks.

The new view is that they are dynamic, non-equilibrium systems that are constantly changing and adapting, and that where organization exists it has emerged spontaneously from the interactions of the component parts. As urban theorist Michael Batty has said, "Planned cities are always the exception rather than the rule and when directly planned, they only remain so for very short periods of time."

The organic quality of urban growth has long been recognized. The American social theorist Lewis Mumford called growing cities "amoeboid", and considered the uncontrolled sprawl of big American cities to be alienating and disempowering, a "crystallization of chaos". But his protégé Jane Jacobs argued that we should trust in the self-organizing vitality of cities rather than in received ideas of what they should look like. In her 1962 book The Death and Life of Great American Cities, she attacked modernist urban renewal schemes as a "mad spree of deceptions and vandalism and waste". Her ideas stimulated the movement known as New Urbanism, which argues that good cities emphasise characteristics such as walkability, diversity, neighbourhood structure and sustainability. Jacobs insisted that cities should be considered as living organisms, with their own metabolism and modes of growth. Her discussion is one of the first in any discipline to acknowledge how complex systems of many interacting parts can display orderly, self-organized behaviour.

Despite these insights, urban planning has been dominated until relatively recently by the view that the city is a kind of machine, comprised of many parts designed to operate together to fulfill a purpose, to meet certain goals or targets. This top-down view of form and function assumes that behaviour can be imposed by fiat: that the components of the urban system will be shaped by a vision of what they should do. It is now apparent that cities do not, in general, live up to this image. When cities succeed, that might often be irrespective of, or even in spite of, planning. When they fail, it is not so much because of bad planning as of an inability to make planning effective at all. Cities have a life of their own, and it is one that arises spontaneously from the interactions of the component parts.

As with other complex systems, this self-organization spawns certain regularities. There are, for example, extremely general mathematical laws that describe how many urban activities, ranging from technological innovation to income, employment levels and power consumption, depend on a city's population size, regardless of its geographical and historical setting. These take the form of power laws, which we have seen previously to be a common signature of complex systems. The quantity in question is proportional to the population size raised to some power, which is typically greater than 1 for quantities reflecting wealth creation and innovation, but less than 1 for quantities that depend on physical infrastruc-

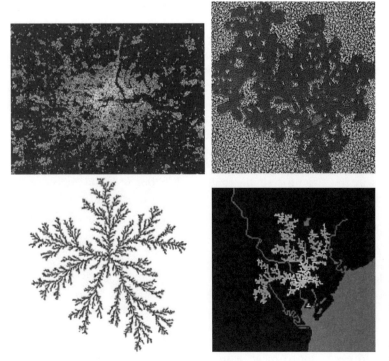

The shapes of cities. A real city (top left) resembles the shape of a cluster of particles grown by aggregation (top right). One common model of such aggregation processes produces ramified, fractal shapes (bottom left), and with a little adaptation it can produce a plausible mimic of a real city confined by coasts and rivers (bottom right). (Credits: (top left & lower right) Courtesy of Michael Batty, University College London; (top right) Courtesy of Arne Skjeltorp, Institute for Energy Technology, Kjeller; (lower left) from D. L. Turcotte & W. I. Newman, *Proc. Natl. Acad. Sci. USA* **93**, 14295 (1996) and copyright National Academy of Sciences, USA.)

ture. One of the implications of this finding is that the pace of urban life increases with a city's physical scale, much as we might expect. Moreover, it seems likely that the self-organization of cities operates in a hierarchical manner, with 'laws' at each level of the hierarchy that do not depend on the fine details of those operating at lower levels.

Some early theories of city growth as a complex system hinted at the origins of universal laws like these. They drew an analogy with the growth of 'fractal' clusters from randomly drifting particles that stick together when they touch: a process called diffusion-limited aggregation that has been used to model the structures of soot, cracks and snowflakes. The analogy here was in the way that new development units, such as business or residential neighbourhoods, are gradually added to the city by 'attaching' to existing development. The cities grown by such models have the ramified and rather disorderly boundaries of real urban centres: densest in the centre and getting more tenuous towards the periphery.

But these fractal cities grow as a single mass, whereas in reality areas of development at the edges of a big city are not always part of the main 'cluster'. There are typically many little satellite towns, which may be swallowed up as the city boundaries sprawl. A better model allows the components of the urban mass to interact: development attracts further development. If two small population clusters grow close to one another, for instance, there is a greater-than-average chance that development will spring up between them: shops to serve the new inhabitants, or local businesses keen to gain a foothold in an up-and-coming area. In other words, the growth of new clusters is interdependent (correlated). When that factor is included, the 'simulated city' that emerges is a clumpy form decorated with sub-clusters and tendrils. This model can also mimic how city shapes evolve over time, and reproduces some of the observed mathematical relationships of urban centres, such as that between number of settlements (cities, towns, villages, hamlets) in an urbanized area and their size.

Universal Maps

Looking at the boundaries of cities tells only part of the story about their growth and form. Urban theorist Bill Hillier has argued that comparisons between cities in many different cultures seem to point to a universal spatial pattern of streets and neighbourhoods which has been called a 'deformed wheel': a centre linked by radial 'spokes' to a surrounding grid of residential areas. This pattern, which can be discerned in cities from Tokyo to Venice to Baltimore, repeats hierarchically at different scales: in local districts as well as in the whole city. By encouraging free flow of pedestrians and traffic, the spokes promote safety, in contrast to the preference of some planners to build cul-de-sacs. Superimposed on this universal structure are culture-specific variations: the complex residential districts of Arabic cities, for example, reflect a stronger separation of public and private life.

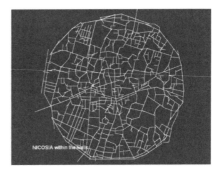

The 'deformed wheel' backbone of city street grids is exemplified here in the city of Nicosia in Cyprus. (Credit: Courtesy of Bill Hillier.)

A mapping scheme in which open spaces are represented as straight lines that are extended until they cross another line, creating so-called axial maps, reveals other aspects of the generic nature of city layouts. The probability distributions of axial line length for many different cities suggest that two distinct types of city structure exist: relatively 'open' structures, with many axial lines that cross the whole urban space (such as Bangkok, Eindhoven, Se-

The shape and growth of a real city like Berlin (left, in 1945) is rather well mimicked by a model of 'correlated' aggregation of particles representing the urban units (right). (Credit: from H. A. Makse, S. Havlin & H. E. Stanley, *Nature* **377**, 608–612 (1995).)

attle and Tokyo), and a denser web of lines dominated by short ones (London, Hong Kong, Athens and Dhaka). The growth of the former group of cities seems to have been influenced by global planning of the large-scale structure, while the latter group has been guided only by local planning, so that there are fewer city-scale features such as long avenues.

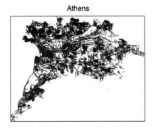

'Axial maps' of cities, in which straight lines designating open spaces are extended until they cross, reveal distinct classes of urban structures, as shown here in examples of open (Tokyo, Bangkok), dense (Athens), and intermediate (Las Vegas) structures. (Credit: from R. Carvalho & A. Penn, *Physica A* **32**, 539–547 (2004).)

Clearly, then, unplanned but statistically predictable structures do emerge through self-organization in cities. Exactly how that happens, and what kind of structures result, are still debated. But the basic principle is not hard to discern: here, as in many natural systems, self-organization is the consequence of interactions between a system's component parts. Business development tends to occur on empty land, which stimulates the construction of new transport and communication infrastructure, attracts service industries, and eventually encourages the development of residential areas. All of these things are interdependent. A road linking two districts may attract development in the area in between. Business and housing attract more business and housing.

These processes can be captured in so-called cellular-automata models, closely related to the agent-based models mentioned earlier, in which the state of each element on a grid is determined by those of the surrounding grid points according to some simple rules. One such model allows five distinct land uses – housing, manufacturing/primary industry, commerce and services, transport in the form of the street/road network, and vacant land. Each land use can generate quantities and locations of other land uses according to certain rules: for example, the development of industrial space might depend on whether there are adequate transport links and available free land, while commercial districts might need to be close to residential ones, and so forth. Land use can also decline, so that cells in the grid become vacant again. Because of lags in redevelopment, urban growth in this model is not necessarily monotonic but can show dips and oscillations, just as has been seen for cities such as London and New York.

Ideally a self-organizing process of city growth will produce a globally optimal structure in terms of, say, the efficiency of land use or transportation. But there is no guarantee that this will be so, and indeed the two solutions will differ in general. The conventional planning approach is simply to impose a (supposedly) optimal configuration by regulation, for example by penalizing sub-optimal land use. But one model of urban self-organization suggested that the emergent structure might instead be guided towards the optimal one by 'pinning' just a few key cells to certain land uses, so that they might act as catalysts that help the dynamic urban pattern converge on the optimum. That is welcome news for planners, because it implies that there is no need to expend lots of resources, or to apply restrictive land-use regulations over large areas of urban space, in order to steer development towards a desired goal. Rather, it may be feasible to identify the crucial 'catalytic sites' that will help a city to evolve spontaneously towards the state that maximizes its efficiency and potential.

This lack of a need for strong central planning supports the contention of the eminent social theorist Herbert Simon that its absence does not necessarily result in poorly 'designed' cities. On the contrary, Simon claimed, they are (or at any rate, they once were) often remarkably effective in arranging for goods to be transported, for land to be apportioned between residential, business and manufacturing districts, and for a lot of activity to be fitted into a small area. Given the right conditions, these things can take care of themselves.

Planning or Managing?

Michael Batty has suggested that the view of cities as dynamic complex systems means that

> *planning, design, control, management – whatever constellation of interventionist perspectives are adopted – are difficult and potentially dangerous. If we assume that social systems and cities [are] like biological systems... then interventions are potentially destructive unless we have a deep understanding of their causal effects. As we have learned more, we become more wary of the effects of such concerted action.*

However, far from being a prescription for stasis or laissez-faire, this situation argues that we need good models of urban complexity that will enable us to try out solutions 'in the laboratory' before we implement them in the real world. It is increasingly important that urban planners be provided with such models in order to pose 'what if' questions. What if we build a high-speed rail link here? What if we impose a congestion charge or reduce bus fares? What if climate change renders this part of the land unusable because of flood risk? What if natural disasters or terrorist attacks necessitate evacuation of part or all of the city?

One model of this sort, developed at the Argonne National Laboratory in collaboration with the US Department of Energy, is called TRANSIMS. It began as a model of urban transportation systems, but has since expanded into a simulation of many aspects of urban development, using census data to construct 'synthetic' populations: individuals and households who have daily activities to which they will need to travel. The TRANSIMS model has been used to investigate transport options, including carbon emissions, for the city and environs of Portland, Oregon. There is a proposal to extend it to Phoenix, Arizona, and the system has also been used to model mass evacuation scenarios from the Chicago metropolitan area. While still primarily focused on one aspect of urban development (transport), TRANSIMS illustrates that such issues cannot be studied in isolation: they require a detailed picture of many facets of urban life: where people live and work, say, and how they choose to go between the two. This approach is likely to become a vital ingredient in the quest to make cities smarter.

Further Reading

H. van Ginkel, 'Urban future', *Nature* **456**, 32–33 (2008).

B. Hillier, *Space is the Machine*. Space Syntax, London, 1996.

M. Batty, *Cities and Complexity: Understanding Cities with Cellular Automata*, Agent-Based Models, and Fractals. MIT Press, Cambridge, Ma., 2005.

Y. Xie & M. Batty, 'Integrated Urban Evolutionary Modeling, in P. M. Atkinson, G. M. Foody, S. E. Darby & F. Wu (eds), *Geodynamics*, pp.273–293. CRC Press, Boca Raton, Fl., 2005.

M. Batty, 'Cities as complex systems: scaling, interactions, networks, dynamics and urban morphologies', *Working Paper* **131**, UCL Centre for Advanced Spatial Analysis (2008).

M. Batty, 'The size, scale, and shape of cities', *Science* **319**, 769–771 (2008).

H. A. Makse, S. Havlin & H. E. Stanley, 'Modelling urban growth patterns', *Nature* **377**, 608–612 (1995).

F. Schweitzer & J. Steinbink, 'Urban cluster growth: analysis and computer simulations of urban aggregations', in F. Schweitzer (ed.), *Self-Organization of Complex Structures: From Individual to Collective Dynamics*, pp.501–518. Gordon & Breach, London, 1997.

R. Carvalho & A. Penn, 'Scaling and universality in the microstructure of urban space', *Physica A* **32**, 539–547 (2004).

F. Semboloni, 'Optimization and control of the urban spatial dynamics', *Complexus* **2**, 195–203 (2004/2005).

Bettencourt, L. M. A., J. Lobo, D. Helbing, C. Kühnert & G. B. West, 'Growth, innovation, scaling, and the pace of life in cities', *Proc. Natl Acad. Sci. USA* **104**, 7301–7306 (2007).

K. Nagel, R. J. Beckman & C. L. Barrett, '*TRANSIMS for Urban Planning*', Paper LA-UR 984389. Los Alamos National Laboratory, Los Alamos, Nm.,1999.

D. Helbing & K. Nagel, 'The physics of traffic and regional development', *Contemp. Phys.* **45**, 405–426 (2004).

Peaks of heat emission predicted for a simulated evacuation of Naperville, Illinois, using the TRANSIMS model of urban transportation. (Credit: Argonne National Laboratory - Transportation Research and Analysis Computing Center.)

The Transformation of War: Modelling Modern Conflict

War is not what it used to be. Since the late twentieth century there has been a discontinuity in the very nature of war, a fact that the rhetoric of a 'war on terror' following the attacks on the World Trade Center on 11 September 2001 failed catastrophically to acknowledge. It is hard to avoid the suspicion that this 'war on terror' was one for which political leaders were determined to find conventional battlegrounds – in Afghanistan and Iraq – only to discover that the battle refused stubbornly to materialize, because that is no longer what armed conflict is about. Typically, there is not in any meaningful sense a declaration of war to kick things off, nor a peace treaty to conclude them. Formal armed forces are peripheral; so are formal leaders. Armies may be mobilized not to fight war but (allegedly) to keep peace – or, in Afghanistan and Iraq, to do both at once. According to strategic analyst Anthony Cordesman, "One of the lessons of modern war is that war can no longer be called war."

The changing character of wars argues for a shift in military tactics. In many ways, fighting against terrorist-style insurgency is like fighting an illness that continually evolves, adapts and changes. Political scientist Mary Kaldor asserts that US military action in Iraq was predicated on the view that it was a war much like those fought until the middle of the twentieth century, where two military states vie for control of a territory. This, she says, is the wrong approach – and what is more, "The US failure to understand the reality in Iraq and the tendency to impose its own view of what war should be like is immensely dangerous." Instead of approaching it as a conflict that can be conclusively won by military force, they should see it as an ongoing effort, Kaldor argues. For terrorist-style conflicts seem to be sustainable indefinitely: "These wars," says Kaldor, "are so much harder to end than to begin."

How and why are these changes happening? Are we seeing a change in the reasons why many (if not all) wars begin, or in the ways they are waged, or in the objectives that are pursued, or in the nature of the combatants? Or all of these, and more?

The Power of War

According to Carl von Clausewitz in his classical 1832 treatise On War, "War is an act of violence to compel our opponent to fulfil our will." But while for Clausewitz a 'theory of war' (he was unsure if this was an art or a science) pertained simply to the best way to wage it, violence is not the only, or necessarily the best, way to compel your opponent. Wars are immensely costly, and put at risk the very existence of the states waging them. A theory of war must surely examine why opponents go to war, and should explain why sometimes they do not.

It was in the hope of answering such questions that, from the 1920s to the 1950s, the British physicist Lewis Fry Richardson gathered statistic data on what he called 'deadly quarrels': armed conflicts ranging from small local skirmishes to world wars. Richardson, a pioneer of meteorology, was a Quaker and conscientious objector in the First World War, and hoped that by analysing the 'facts of war' he could find a way to promote world peace. He found that his data could be described by a surprisingly simple mathematical relationship: the number of conflicts with N fatalities declined such that the probability of such a conflict was proportional to $N^{-\alpha}$ where α is called the exponent. This form, which we have encountered several times earlier in the book, is called a power law. It implies that small wars are more common than large ones; specifically, a tenfold increase in the severity of a war decreases its probability by a factor of 2.6.

Similar power-law relationships between size and frequency apply to other phenomena, such as earthquakes and fluctuations in economic markets. The persistence of the law for events of all sizes indicates that even the biggest, most infrequent earthquakes are created by the same processes that produce hordes of tiny ones, and that occasional market crashes are generated by the same internal dynamics of the marketplace that produce daily wobbles in stock prices. In the context of warfare, it suggests that there may be a degree of universality in the mechanisms,

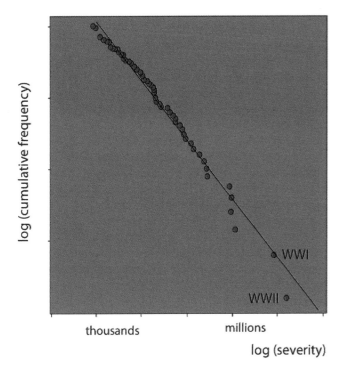

How the number of deaths in a war between states (from 1820 to 1997) varies with the frequency (equivalently, the probability) of such a conflict. The relationship is a mathematically simple one, called a power law. The last two data points are, respectively, World Wars I and II.

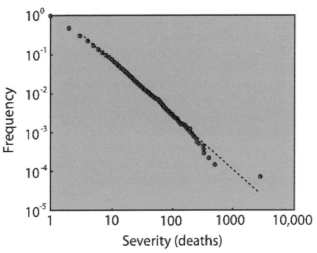

How the number of deaths in a terrorist attack (between 1968 and 2008) varies with the frequency of such an attack. The relationship is another power law, but with a different exponent from that of inter-state wars.

of war: small ones are qualitatively the same type of phenomenon as big ones.

What's more, the power-law relationship implies that one cannot tell how big any particular conflict will become, just as one cannot tell whether an earthquake or landslide, once triggered, will be a major or a minor event. Perhaps a war will destabilize neighbouring regions and spread like a forest fire; or perhaps it will remain localized. Which of these will be the case seems almost impossible to predict. The implication is that, at least during the nineteenth and twentieth centuries, international relations existed in a state that was prone to this propagation of conflict – not inevitably, but with a potentiality that made interstate concord extremely fragile.

The same kind of power-law relationship can be seen in the statistics of terrorist attacks: that is, in a graph of the number of attacks plotted against their severity (in terms of injuries and/or fatalities). This relationship holds for events ranging from those that injured or killed just a few people to those that, like the Al Qaeda Nairobi car bomb in August 1998, produced over 5,000 casualties.

As with wars, the power-law relationship implies that the biggest terrorist attacks such as 9/11 are not 'outliers', one-off events somehow different from suicide bombings that kill or maim just a few people. Instead, it suggests that they are somehow driven by the same underlying mechanism. Note, however, that the precise form of the power law depends on the type of country to which it relates. Terrorist attacks in Western industrialized nations are rare but tend to be large when they happen. Terrorist attacks in the less-industrialized world tend to be smaller, more frequent events. The two classes are distinguished by different power-law exponents, which reveal some qualitative distinction that is yet to be understood.

Similarly, the slope of the power-law graph – the exponent α of the power-law – for conventional wars differs from those for terrorist attacks, which suggests that the two are also distinct classes of conflict: terrorism is not like conventional war. However, these distinctions are becoming far less evident in modern warfare. The recent war in Iraq and the ongoing civil conflict in Colombia at first followed the Richardson power-law for traditional warfare, but later approached the different power law characteristic of terrorist attacks in non-industrialized nations. In Iraq, the conflict began as a conventional confrontation between large armies. But smaller attack units came to predominate as the war continued. Between 2003 and 2005, the 'casualties per attack event' for Iraq followed a gradually changing power-law which came more closely to resemble that found for terrorism. The same trend is seen in the Colombian conflict, which has been fought between the government and various left- and right-wing guerrilla groups for many decades.

These findings are preliminary, and still debated. Some civil wars still look like conventional ones, while in other cases (such as the Algerian civil war of the 1950s and 60s) the blurring of conventional and terrorist-style conflict goes back a long way. Perhaps the point is not so much that all 'new' wars are different from all 'old' ones – that is almost certainly not the case – but that we can hope to use the tools and concepts of complexity to identify some distinctions between them.

Computerized Conflict

The key question is then surely: why do these differences (sometimes) exist? In honesty, no one knows. One idea is that terrorist attacks emerge from a balance between two factors. The severity of an attack is likely to increase in relation to the time spent planning it. But the longer this timescale, the more likely it is that counter-terrorism measures will intervene and suppress the attack before it happens. This balance can, in certain circumstances, create a power law.

Another model tries to understand how terrorism and insurgency evolve from the bottom up: how terrorist networks composed of 'cells' of different sizes arise. The idea is based on emerging understanding of how terrorist organizations are structured: unlike armies, they tend to be decentralized, operating in loose-knit networks that are to some extent autonomous. Moreover, they draw their members from communities that include radicalized individuals who may become inclined towards terrorist activities. Cells can accumulate new individuals, but they also grow by aligning with other cells – and by the same token they may fall apart, for example because of internal conflicts. With the final ingredient that an attack launched by a particular cell has a severity proportional to its size, this 'aggregation' model of terrorist organizations can generate power-law casualty statistics with essentially the same exponent as that seen in the real data.

These models are evidently not the last word; in many ways they are simplistic and crude. But they illustrate that the statistics of armed conflicts demand new ways of looking at how they arise, and that agent-based models predicated on interactions between the key players can at least in principle offer explanations of why these numbers take the forms that they do. Most importantly, they can be used to address questions such as which counter-terrorism strategies might be effective: whether it is better, for example, to focus on reducing the radicalized proportion of a population, or to seek out the largest cells, or to sever links between cells.

A perspective based on cost-benefit choices made by aggressors can also be productive in understanding how and why terrorism arises in the first place. Despite the moral rhetoric of nations and governments that have suffered terrorist acts, terrorism is not a nihilistic gesture but a strategic choice – a particular mode of combat usually deployed by groups that lack military might, and which may (as with the ANC in South Africa or the Republicans in Northern Ireland) be abandoned if more advantageous options are available. The dynamics of terrorist activities may be different when there is just one, or several competing parties involved in them, such as Fatah and Hamas in Palestine. However deplorable it might seem, terrorism is rarely irrational. All the same, the statistics of terrorist attacks in the Isareli-Palestinian conflict suggest that the causative factors are considerably more complex than current theories tend to predict.

Why Fight?

This comes back to the broader question posed at the outset: why do conflicts happen? The issue has naturally received a huge amount of attention in recent decades, whether from military strategists, political scientists or agencies devoted to fostering peace. But while no one imagines that there will ever be a one-size-fits-all theory of warfare and conflict, it's possible to identify some general questions that might have quite general answers.

For example, it is obvious that many wars between states happen when they share a border or a resource such as water. There is a clear link to economic prosperity: in the modern era, wealthy nations do not tend to fight one another. The actors in many of the most violent modern conflicts have been defined in terms of ethnicity and culture rather than nationality: for example, in former Yugoslavia, Rwanda and the Democratic Republic of the Congo. Why do some ethnic groups seem able to coexist stably with one another, while others do not? What factors seem to spark *civil* wars? In particular, and contrary to some political naivety about interventions and uprisings in Iraq, Afghanistan and North Africa, the process of democratization seems often to make internal conflict *more* likely (even if the achievement of a stable democracy does eventually foster excellent prospects for subsequent national and international peace).

Traditional political science has tended to regard states as actors with fixed borders that engage in self-interested competition, based primarily on national interest and security: the so-called realist paradigm, which accords with Clausewitz's view that war is merely a continuation of politics by other means. Here, international relations represents a struggle for power much like that

A snapshot from the GeoSim model of nationalist insurgency in a model state with 'real' topography. Rebelling provinces are indicated as red needles, and are located mostly in mountainous areas. (Credit: from L.-E. Cederman & L. Girardin, paper prepared for the Annual Meeting of the Am. Polit. Sci. Assoc., 2007.)

invoked between individuals by Thomas Hobbes in the seventeenth century.

This picture lends itself well to agent-based modeling. One such model, called GeoSim, developed in the late 1990s, has been used to explore the effects on interstate conflict of such factors as alliance formation and democratization. The GeoSim model has furnished a possible explanation of the power-law statistics of wars, based on the way technological change alters the balance of strengths between states and thus (in the realist view) their readiness to wage war.

Such models can be adapted to explore broader questions. GeoSim has, for example, been used to look at civil wars in which provinces rebel against the central state authority. Here geography proves to be important: violence is often concentrated in mountainous regions, which are hard to suppress. There is now a strong perceived need to tie such models more explicitly to real data and to allow them to explore the many dimensions of conflicts, such as onset, duration, spatial extent and casualty levels.

Civil wars are in fact a key focus of current research on conflict, not least because they are so widespread. During the past several decades, civil war was been waged on average in 1 out of every 10 countries worldwide, a disproportionate number of them being poor. These conflicts challenge the traditional view of a state as being 'at war' or not, since such conflicts are often localized – in Kashmir or Chechnya, say. As a result, they lend themselves to – indeed, they demand – a picture that considers interactions between several different actors, influenced by complex, local and heterogeneous factors within a state. In one recent model, for example, either (two-sided) civil violence or (one-sided) government repression was found to arise in a two-faction population in response to fluctuations in wages or aid, if political institutions were weak. This is consistent with the observation that civil war is highly sensitive to economic factors, rather than primarily to social or cultural ones: diverse populations can coexist peaceably if they are wealthy enough.

The more we understand these issues, the more they suggest that violent conflict fits within the broader picture of complex social systems. For example, the likelihood of civil violence depends on the specific modes of organization in rebel groups. Large excluded groups have more resources, but may not have much coercive power if, as with the Palestinians, they are fragmented into several competing organizations. Small, cohesive organizations have a disproportionate tendency to fight. This makes it important that models include some representation of spatial and social network structures and communities, rather than just undifferentiated hordes of aggressors.

One of the big questions about the onset of violence within a region, state or population is whether it is rendered more or less likely by cultural or ethnic segregation. Anecdotally, either possibility can be attested. There is good evidence that social desegregation improves tolerance and reduces hostility, while segregation can harden prejudice, as seen for example in Northern Ireland or the racially segregated cities of northern England. On the other hand, the genocide in Rwanda in 1994 took place in the context of a highly mixed population of Tutsi and Hutu people, while urban violence between Muslims and Hindus in Ahmedabad in 2002 was greater in mixed rather than segregated neighbourhoods. Ethnic migration – a lessening of mixing – following deadly attacks has been found to reduce violence.

These apparent contradictions can only be understood by considering the spatial element of conflict: where the actors physically reside in space. One such model was based on the segregation model of Thomas Schelling (see page 14) but with an added impetus for migration: a wish to escape recent violent conflict in the neighbourhood. Interactions between the agents were represented by a concept of 'social distance', which depended on several factors: the greater the distance, the greater the tension, and above a certain threshold this tension could erupt into violence. The model was used to simulate the situations in Jerusalem during the violent *Intifada* of 2001–4 and the somewhat more settled period of 2005–9, based on real data about the spatial demography of the Muslim and Jewish populations at these times. Having established that the model did a reasonably good job of predicting the spatial distribution and severity of outbreaks of violence during both periods, the researchers used it to explore the likely consequences of different real-world proposals for distributing the populations: from complete mixing to different modes of segregation defined by the division of city districts into those under Israeli or Palestinian authority. Here the fully mixed scenario produced the most violence. But while it declined in segregated scenarios, there was less violence when segregation was partial (most markedly for the case of a return to 1967 boundaries) than when it was total.

Models like this are certainly not advanced enough yet to offer strong policy recommendations. But they teach an important lesson: outcomes of interventions in situations like this are not always intuitively obvious, since they involve a complex interplay of effects. And simplistic notions that segregation is universally 'good' or 'bad' for reducing violent conflict are potentially misleading: it is unwise to assume that 'one size fits all'. While in this model the primary driver of unrest is 'social distance', the factors contributing to this measure will differ from place to place – they may be economic, cultural, or linked to property rights, say. Models like this can be tuned to the local specifics of culture and geography, rather than purporting to make general statements about what does and does not work.

As all these examples illustrate, there is diversity in the causes of conflicts, but also some regularity and perhaps even predictability. War and other violent conflicts are complex phenomena, but not random ones. It is far from utopian to imagine that they can be understood, at least in part, as a complex social phenomenon which might therefore be amenable to planning, guidance, control and mitigation.

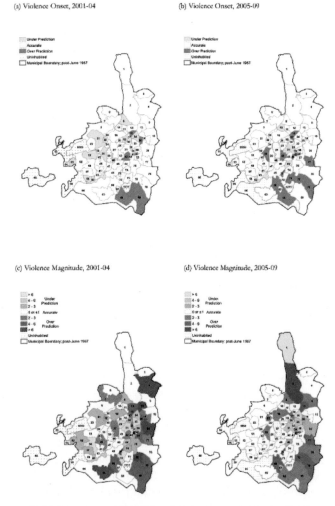

Predictions of a model that describes tensions and violent outbreaks in Jerusalem as a function of the segregation of Jewish and Muslim populations. Here the colour coding shows how the predicted levels of violence compare with those in reality over two time periods. (Credit: from R. Bhavnani et al., submitted (2012).)

Further Reading

A. Clauset, M. Young & K. S. Gleditsch, 'On the frequency of severe terrorist events', *J. Conflict. Resolut.* **51**, 58–87 (2007).

A. Clauset & F. W. Wiegel, '*A generalized aggregation-disintegration model for the frequency of severe terrorist attacks*', preprint http://arxiv.org/abs/physics/0902.0724 (2010).

N. Johnson et al., '*Universal patterns underlying ongoing wars and terrorism*', preprint http://arxiv.org/abs/physics/0605035 (2006).

L.-E. Cederman, 'Modelling the size of wars: from billiard balls to sandpiles', *Am. Polit. Sci. Rev.* **97**, 135–150 (2003).

J. C. Bohorquez, S. Gourley, A. R. Dixon, M. Spagat & N. F. Johnson, 'Common ecology quantifies human insurgency', *Nature* **462**, 911–914 (2009).

R. Bhavnani, D. Miodownik, K. Donnay, M. Mor & D. Helbing, '*Group segregation and urban violence*', submitted (2012).

L.-E. Cederman & L. Girardin, '*Towards realistic computational models of civil wars*', paper prepared for the Ann. Mtg of the Am. Polit. Sci. Assoc., 2007.

L.-E. Cederman, S. Hug & A. Wenger, 'Democratization and war in political science', in *Democratization,* eds W. Merkel & S. Grimm 15, 509–524 Routledge, London, 2008.

H. Strachan, '*The changing character of war*', Europaeum Lecture, Geneva, 9 November 2006.

Summary

For the past 30 years or so there has been a more or less conscious effort both to make societies more globalized and to foster technological advances that promote social interdependence. But we have not made a concomitant effort to build a science that enables us to understand the strongly coupled systems we have created. To do so, we need to reach out beyond conventional disciplinary borders and to develop tools for exploring the interactions of many aspects of society. This book has argued that, as a result, it is valuable and necessary to consider society as a whole as a complex system that can be explored, explained and to some extent predicted using idealized models.

In most of the studies conducted so far, these models have been highly simplified. In some cases, however, such as the descriptions of traffic, pedestrian movement and social networks, they are already adequate to address real-world problems and needs. And experience with these attempts to understand social behaviour, as well as with complex systems more generally in the physical and biological sciences, offers some important messages about how we might best manage our increasingly complicated and interconnected human world. For example:
- Abandon linear and one-size-fits-all thinking.
- Don't impose solutions, but create the conditions for effective solutions to emerge spontaneously.
- Let randomness rescue you from dead ends, bottlenecks and non-optimal states.
- Embrace and take advantage of diversity.
- Let self-organization create adaptability and innovation.
- Remember that where you are may depend on how you got there: history matters.

One key objective of this book is to show that, while some social phenomena can be studied in isolation, in general the challenges we face pertain to many strongly connected social activities, which are often themselves coupled to processes in the natural world. To understand the spread of infectious diseases, for example, we need to know about infection, recovery and death rates, but also about how people come into contact with one another and how they move around, whether in routine trips or via trans-continental travel. This in turn requires a consideration of, among other things, public transportation and social activity patterns, including behavioural changes in response to epidemic spreading. Evidently, this picture can become extremely complex, and exceeds the expertise of any single academic discipline, including traditional epidemiology itself.

But this does not mean that the situation is hopeless or intractable. Many of these issues are already reasonably well understood in isolation. Others, such as the patterns of human mobility, are new fields of study but have benefitted from major advances in conceptual understanding and empirical data-collection. What is now needed is a means of integrating these various topics so that some of the most serious challenges we face on the planet – war, poverty, economic instability, environmental change – can be studied in a flexible yet comprehensive manner.

Climate science can serve as a good analogy here. The field has steadily expanded its scope beyond a consideration of how greenhouse-gas emissions will affect the energy balance of the atmosphere, to include such factors as the cycling of elements like carbon and nitrogen between living and inorganic reservoirs in the oceans and on land, the dynamics of ice-sheet growth and movement, the circulation patterns of the deep and shallow oceans, and now increasingly the dependence of greenhouse-gas emissions on human economic and social activity. Climate researchers understand that these developments will not and should not result in a single, over-arching computer model on which all forecasts will be based; rather, what is needed is a suite of complementary models that together offer multiple perspectives of the range of possible futures and their relative probabilities and uncertainties. We might recognize too that despite ongoing controversy about the long-term predictive capabilities of climate models, in the shorter term modelling the same suite of physical phenomena – atmospheric and ocean dynamics, cloud formation and precipitation, effects of ecosystems and so forth – has furnished a largely reliable and trusted tool for weather prediction that is now routinely used for decision-making.

It is time to build a similar broad platform for modelling and prediction of many other social and environmental challenges, using the approaches, concepts and tools described in this book. Such an effort would require large-scale federated coordination, and would have to promote interdisciplinary integration of natural, social, and engineering sciences to address a wide range of global challenges. Of course, we cannot expect models to solve all our problems or reliably to predict our future in detail. However, again weather forecasting demonstrates that even somewhat approximate and uncertain short-term predictions can be useful, for agriculture, for air traffic, and for everyone. Such forecasts have improved considerably over time, by combining models with supercomputing and real-time sensing. The investment in such systems today is thought to repay itself many times in the benefits it confers. Even though humans are doubtless much harder to model and predict than the physical processes that underlie the weather, one can certainly anticipate similar benefits from modelling our techno-socioeconomic systems. This is an endeavour whose time has arrived.

New Ways to Promote Sustainability and Social Well-Being in a Complex, Strongly Interdependent World: The FuturICT Approach

Dirk Helbing

FuturICT is one of six proposals currently being considered for support within the European Commission's Flagship Initiative (see Box 1). The vision of the FuturICT project is to develop new science and new information and communication systems that will promote social self-organization, self-regulation, well-being, sustainability, and resilience. One of the main aims of the approach is to increase individual opportunities for social, economic and political participation, combined with the creation of collective awareness of the impact that human actions have on our world. This requires us to mine large datasets ('Big Data') and to develop new methods and tools: a Planetary Nervous System (PNS) to answer "What is (the state of the world)..." questions, a Living Earth Simulator (LES) to study "What ... if..." scenarios, and a Global Participatory Platform (GPP) for social exploration and interaction.

Today, we understand our physical universe better than our society and economy. Challenges like the financial crisis, the Arab spring revolutions, global flu pandemics, terrorist networks, and cybercrime are all manifestations of our highly and ever more connected world. They also demonstrate the gaps in our present understanding of techno-socio-economic-environmental systems.

In fact, the pace of global and technological change, in particular in the area of Information and Communication Technologies (ICTs), currently outstrips our capacity to handle them. To understand and manage the dynamics of such systems, we need a new kind of science and novel socially interactive ICTs, fostering transparency, trust, sustainability, resilience, respect for individual rights, and opportunities for participation in political and economic processes.

Information and Communication Technologies (ICTs) are increasing playing a key role for the understanding and solution of problems that our society is facing. Many ICT devices take autonomous decisions, based on real-world data, an internal representation of the outside world, and expectations regarding the future. In some sense, ICT systems are increasingly becoming something like *'Artificial Social Systems'*. Already today, supercomputers perform most financial transactions in the world.

The FET Flagship Initiative

The FuturICT project is the response to the Flagship Initiative launched by the Future and Emerging Technology (FET) section of the European Commission: a call for 'Big Science' projects with genuinely transformative potential and a 'man on the moon' scope of vision. In the first round of the process, 21 candidates were narrowed down to six Flagship Pilots, of which FuturICT received the highest rating.

Each pilot will submit detailed proposals in April 2012, and at least two of them will be selected for funding of up to €1 billion each over ten years, starting by the end of 2013. (Note that this is about a tenth or less of what is currently invested in other Big Science projects such as the Large Hadron Collider at CERN, the ITER nuclear-fusion reactor, the Galileo satellite program and the Human Genome Project.) Approximately half of the money must be mobilized by the project partners from national budgets and funding agencies, from business and industry, or from donations. A considerable fraction of the Flagship funding will be distributed through 'Open Calls', which will allow a wide scientific community to contribute to the goals.

Among the particular strengths of the FuturICT proposal are:
- its societal relevance,
- the immediate importance of its results for the everyday life of ordinary citizens,
- its large and rapidly growing community and multidisciplinary nature,
- the participation of many European countries,
- the significant support of scientific communities in other continents,
- its open-access and transparent project architecture, and
- its educational activities.

However, today's ICT systems are not constructed in a way that ensures beneficial outcomes. This can result in problems, which we are also facing in our real society, e. g. breakdowns of coordination and performance, 'tragedies of the commons', instabilities, conflicts, (cyber) crime, or (cyber-)war. Furthermore, ICT systems influence not just their own state, but also impact the real world and human behavior. We, therefore, need a deep understanding of techno-social systems to get ICT systems right and also mitigate our societal problems.

We now have a global exchange of people, goods, money, information, and ideas, which has created a strongly coupled and strongly interdependent world. This often causes feedback and cascading effects, extreme events, and unwanted side effects. In fact, these systems behave fundamentally differently from weakly coupled systems. Multi-component systems can be dynamically complex and hard to control. Therefore, we need a paradigm shift in our thinking, moving our attention from the properties of the system components to the collective behavior and emergent systemic properties resulting from the interactions of these components.

The paradigm shift from a geocentric to a heliocentric worldview has facilitated many things, from modern physics to our ability to launch satellites. In a similar way will the paradigm shift towards an interaction-based, systemic perspective and a co-evolution of ICT with society open up entirely new solutions to address old and new problems, such as financial crises, social and political instabilities, global environmental change, organized crime, the quick spreading of new diseases, and how to build future cities and smart energy systems.

As the previous chapters by Philip Ball have shown, there are promising new approaches to manage complexity: While external control of complex systems is hardly possible due to their self-organized dynamics, one can promote a favorable self-organization by modifying the interaction rules and institutional settings. The potential of a flexible, self-regulating approach has been impressively demonstrated for urban traffic light control and a number of other problems. This approach is based on real-time sensing, short-term anticipation, and the implementation of suitable adaptive interaction rules between the connected system elements. The decentralized self-regulatory principle can be scaled up to systems of almost any size and any kind.

To successfully transfer this approach to other areas and make an effective contribution to mitigating our 21st century problems, we need to develop a better, holistic understanding of the global, strongly coupled and interdependent, dynamically complex systems that humans have created. For this, it is necessary to push complexity science towards practical applicability, to invent a novel data science (which reveals how information is transformed into knowledge and influences human action), to create a new generation of socially interactive, adaptive ICT systems, and to develop entirely new approaches for systemic risk assessment and integrated risk management.

The FuturICT project is a perfect opportunity to foster the creation of such knowledge and the development of the fundaments of new information and communication systems such as a 'Planetary Nervous System' to enable collective, ICT-based awareness of the state of our world, a 'Living Earth Simulator' to explore side effects and opportunities of human decisions, a 'Global Participatory Platform' to create opportunities for social, economic and political participation, an 'Open Data Platform' (a 'Data Commons') to foster the creativity of people and new business opportunities, a 'Trustable Web' to support safer, privacy-respecting information exchange, as well as value-sensitive ICT to promote responsible interaction. In fact, Europe could well be leading the upcoming age of social and socially inspired innovations, which comes with enormous societal and economic potential.

Why FuturICT is Needed

The current lack of a project of this scope and ambition is surprising and, one might even say, deplorable. Big Science projects such as the Human Genome Project, the Large Hadron Collider and the Hubble Space Telescope have revealed, or are revealing, fundamental insights into our genetic constitution and the laws that govern our physical world. However, we have not given the social sciences the same priority as the natural sciences, even though they are highly relevant for maintaining and increasing social well-being. As we have seen in earlier chapters, this is partly because traditional approaches to social problems such as violent conflict and economic instability have often not been very effective in alleviating them. But this is due to the fact that societal challenges have a particular, complex nature as a result of the strongly interconnected patterns and structures of life. Strongly connected, dynamical systems have a number of characteristic properties, for example:

- Even the most powerful computers cannot perform an optimization of the system behaviour in real time, when the number of interacting system elements is large.
- Most real-life complex systems behave probabilistically rather than deterministically, i. e. their behaviour cannot be exactly predicted.
- Strongly connected systems with positive feedbacks tend to change fast, often faster than we can responds and collect enough experience about their behaviour.
- Extreme events occur more often than expected, and can impact the whole system.
- Self-organization and strong correlations dominate the system behaviour. This can lead to surprising, 'emergent' properties of the system.

- The system behaviour can be rich, complex and hard to predict. Planning for the future may be difficult or impossible.
- Complex systems may appear uncontrollable. In particular, opportunities for external, top-down control are very limited.
- Due to possible cascading effects, the vulnerability to random failures or external shocks may be great.
- The loss of predictability and control may lead to an erosion of trust in private and public institutions, which in turn can lead to social, political, or economic destabilization.

Some of these properties challenge our common way of thinking and defy intuition. Even in purely financial terms, the consequences of failing to appreciate and manage these characteristics of global systems and problems are immense. For example:

- The financial crisis has caused estimated losses of $20 trillion.
- Crime and corruption consumes 2–5% of global GDP: about $2 trillion annually.
- Global military expenditures amount to $1.5 trillion annually.
- The 9/11 terrorist attacks on the US cost the country's economy $90 billion.
- A true influenza pandemic infecting 1% of the world population would cause losses of $1–2 trillion per annum.
- Traffic congestion costs the economy £7–8 billion in the UK alone.

If FuturICT could reduce the impact of these societal problems by just 1%, this would already represent a return of many times the prospective €1 bn Flagship investment. Based on previous success stories regarding a better management of complex systems, an improved understanding of the fundamental underlying issues could actually be expected to facilitate improvements (e. g., in efficiency) of 10–30%.

There are also strong ethical arguments to support the FuturICT project (see Box 2). The fragility of the financial and economic system, for example, carries a serious risk of endangering the stability of society, which may promote crime, corruption, violence, riots, and political extremism, and ultimately undermine democracies and destroy cultural heritage. Rapid scientific progress is needed to learn how to prevent such cascading effects and deterioration. It is also vital to ensure that the social innovations that a project like FuturICT could engender will benefit all of humanity and not end up in the hands of a few stakeholders – a situation that has threatened to arise, for example, in genetic engineering and other transformative technologies.

Why Information and Communication Technology (ICT) is Crucial

In the global challenges we face, information and communication technologies are part of the problem. People feel that they have created too much speed, too much data,

---

The FuturICT Approach

FuturICT has also a strong ethical motivation. Among its aims are

- to promote human well-being and responsible behaviour,
- to promote the provision of unbiased, high-quality information, and to increase individual and collective awareness of the impact of human behaviour,
- to reduce vulnerability and risk, increase resilience, and reduce damages,
- to develop contingency plans and explore options for future opportunities and challenges,
- to increase sustainability,
- to facilitate flexible adaptation,
- to promote fairness and happiness,
- to protect and increase social capital,
- to improve opportunities for economic, political, and social participation,
- to find a good balance between central and decentralised (global and local) control,
- to protect privacy and other human rights, pluralism and socio-bio-diversity,
- to support collaborative forms of competition ('co-opetition') and
- to promote responsible behaviour.

Over its ten-year funding period, FuturICT will aim to develop a new ICT paradigm, focusing on socio-inspired ICT, the design of a 'trustable web', ethical, value-sensitive, culturally fitting ICT (responsive+responsible), privacy-respecting data-mining technologies that give users control over their own data, platforms for collective awareness, a new information ecosystem, the co-evolution of ICT with society, public involvement and democratic control. FuturICT plans to provide an open data, simulation, exploration and participatory platform to promote new opportunities for everyone. This platform would represent a new public good on which all kinds of services can be built. It will support both commercial and non-profit activities. To prevent misuse and enable reliable high-quality services, the platform will be decentralized and built on principles of transparency, reputation and self-regulation.

and too much complexity. In fact, our global ICT system is the most complex artefact ever created. It is made up of billions of interacting elements (such as computers, smartphones, users, companies, cars, etc.).

Although humans have built the individual components that compose the system, we are increasingly losing the ability to understand the system as a whole and its interaction with society. No one planned this system as a whole, and no one is in control of it. Often, we do not even know what it 'looks' like – what, for example, the topology of the connectivity network is. We also do not know its weak points and vulnerabilities. We can't predict its behaviour.

But information and communication technologies will inevitably also have to be a part of the solution. We've seen earlier how such systems already collect and embody a tremendous amount of data, some of which encodes vital information about performance, reliability and robustness. Moreover, the pervasiveness of such systems creates an infrastructure that allows us to capture the kind of data, which are needed to model and understand our complex techno-socio-economic-environmental systems. It is an opportunity that must be handled with care (see Box 2).

Despite the need for more data, it's important to collect and use it at the appropriate level. That's to say, FuturICT would not need or desire 'all the data in the world', and the objective is not to model every individual in detail. It is in the very essence of complex systems that this level of detail is not needed: many of the important behaviours, such as trends, norms and cultural shifts, are collective ones, which can be understood without knowing everything about the single individuals and their interactions. That is precisely what makes the complexity approach tractable: it does not involve a 1:1 mapping of the world onto models. In general, different questions require one to focus on different levels of hierarchically organized systems or strongly connected parts of a system, with a degree of data aggregation that is fit to the purpose. (Doctors do not need genetic information to fix a tooth, nor brain scans to operate on the knee.)

Concomitant with the need for massive data collection, FuturICT will require innovations in the extraction of information and meaning. Some new technologies now supply means of gathering data in volumes and rates that exceed our ability to store, retrieve, catalogue, and interpret them. This is a problem felt particularly keenly in the science of genomics and bio-informatics, where data collection has sometimes proceeded apace in the absence of a conceptual framework for asking questions and testing hypotheses. Data is good only to the extent that it can be mined for meaning. To generate valuable knowledge, data mining must often be combined with theoretical models. Advances in this area will, therefore, require input from computer scientists and specialists in data visualization. But the perspective of the social and complexity sciences to identify meaningful trends and correlations will also be essential.

The Components of FuturICT

The FuturICT project primarily aims at bringing data, models and people together. It will develop new information and communication technologies (ICTs) to collect massive data sets and mine them for useful or meaningful information, and build ICT systems that have the capacity to self-organize and adapt to the collective needs of users. These ICT systems will be the basis of the *FuturICT Platform*, which will have three main components: the *Planetary Nervous System*, the *Living Earth Simulator*, and the *Global Participatory Platform*. The measurement, modelling, and participatory elements of these components will be used for practical applications, such as *Exploratories for Society, Economy, Technology*, and the *Environment*. These Exploratories serve to interactively explore interdependencies in our world and will be created by connecting *Interactive Observatories* for social well-being, for conflicts and wars, for financial systems, transportation and logistic systems, any many other areas.

The *Living Earth Simulator* will enable the exploration of possible future scenarios at different degrees of detail, employing a variety of perspectives and methods (such as sophisticated agent-based simulations and multi-level models). It will act as a Policy Simulator or Policy Wind Tunnel, enabling one to test alternative choices and different policies in advance to explore their possible or likely consequences. These simulations will be enabled by the so-called *World of Modelling* – an open-software platform, comparable to an App store, to which scientists and developers can upload theoretically informed and empirically validated modelling components that map parts of our real world. Rather than giving an ultimate answer, the pluralistic approach of this platform will offer multiple perspectives on difficult problems and, thereby, support better informed decision-making guided by the values and priorities of the respective users.

The Living Earth Simulator will require the development of interactive, decentralized, scalable computing infrastructures, coupled with the access to Big Data. Gathering these data is the role of the *Planetary Nervous System (PNS)*. Our bodies are constantly responding and adjusting to new data – for example, tuning our metabolism to the energy requirements, ambient temperature and current energy reserves. Sensory feedback is essential for navigating our environment, avoiding danger and performing fine motor tasks. The Planetary Nervous System will provide the same sort of function for our Earth. It will be comprised of a global sensor network, where 'sensors' include anything able to provide data in real-time about socio-economic, environmental or technological systems, including the Internet. Such an infrastructure will enable real-time data mining ('reality mining'), e. g. of news, information feeds, or search trends.

A crucial part of the FuturICT platform are the *Interactive Observatories*, which will use the Planetary Nervous System to spot potential weaknesses or problems arising in specific sectors, such as the financial system or conflict-prone regions. These Observatories will continuously monitor the 'health' of the economy, of urban and transportation systems, or of international relations, say, to give early warnings of hazard, where possible. Analogous to seismic-monitoring networks, they will watch out for the build-up of stresses or for precursory signals of impending collapse or catastrophe. They will thus facilitate pre-emptive searches for solutions based on specific, locally tuned scenarios.

The *Global Participatory Platform (GPP)* will make FuturICT's new methods and tools available for everyone (with reputation and transparency mechanisms in place to foster responsible use). This will enable people to look at all interesting issues from many angles and to use the power of crowd sourcing and the wisdom of crowds. The Global Participatory Platform will promote communication, coordination, cooperation and the social, economic and political participation of citizens. In this way, the traditional separation between users and providers, or customers and producers will be overcome, thereby unleashing new economic potentials. Building on the successful principles of Wikipedia and the Web x.0, societies will be able to harness the knowledge and creativity of many minds much more effectively than we can do today.

The Global Participatory Platform will include *Interactive Virtual Worlds*, where possible futures can be explored and enacted using techniques like those developed for multi-player online games. The purpose of these virtual copies of our world is to explore possible futures, i. e. to identify likely systemic outcomes of interactions, given certain 'rules of the game' and institutional settings. For example, one could study different kinds of financial architectures and the market dynamics resulting from them. Such participatory experiments could also inform the designs of shopping malls, airports, and future cities.

In addition to the interconnected systems forming the Living Earth Platform, FuturICT will create an *Innovation Accelerator (IA)* to identify inventions and innovations early on, to distil valuable knowledge from a flood of information, to find the best experts for projects, and to fuel distributed knowledge generation by 'crowd-sourcing'. This Innovation Accelerator will also catalyse the integration of project activities in one single platform. Because the Innovation Accelerator will support communication and flexible coordination in large-scale projects, it will form the basis of the innovative management of the FuturICT Flagship itself.

Towards More Resilient and Sustainable Systems: How It All Comes Together

One of the scientific challenges behind attempts to promote social well-being will be to measure the relevant factors for it on a global scale and in real-time with sufficient accuracy. (Note that delayed policy response may cause an unstable system dynamics).

Measuring social well-being is even more difficult than measuring GDP. Currently, reliable official numbers for GDP are published only with a delay of many months. However, new ways of measuring GDP have recently been suggested. For example, it seems feasible to estimate GDP from satellite pictures of global light emissions. Such estimates are possible almost in real-time. Similarly, it has been shown that health-related indicators (such as the number of patients during flu pandemics) can be well estimated based on Google Trends data.

Therefore, the vision of FuturICT is to make the different dimensions of social well-being globally measurable in real-time. This could be done by mining freely available data on the internet, by sentiment analysis of tweets and blogs, or by use of sensor data of various kinds. Recent attempts to measure happiness and its variation in space, time and across social communities point the way for this.

Determining the 'Social Footprint' to Protect the Fabric of Society

FuturICT's Planetary Nervous System will provide the methods and tools to measure human activities and socially relevant variables in real-time, on a global scale, and in a privacy-respecting way. By extending measurements to social and economic domains, FuturICT will complement and go beyond the scope of similar projects focused on environmental and climate-oriented measurements (e. g. 'Planetary Skin' and 'Digital Earth'). The ambition of the Planetary Nervous System is more than measurement. It also intends to create individual and collective awareness of the impact of human decisions and actions, particularly on the social fabric on which our society is built (the 'social footprint').

The Planetary Nervous System serves to detect possible opportunities and threats, in order avoid mistakes that one may regret later on. This requires a certain ability to anticipate (in a probabilistic way) possible courses of events. While our own consciousness performs such anticipation by 'mental simulation', FuturICT's Living Earth Simulator will perform the equivalent task for complex techno-socio-economic-environmental systems, by simulating simplified models of our society and economy and other relevant activities in our world. On top of this, Interactive Virtual Worlds and Mixed Reality Environments will provide an online laboratory to explore human interactions under close-to-realistic conditions.

While FuturICT's new concepts for the measurement, simulation and interactive exploration of the impact of human decisions and actions will support collective awareness, the Global Participatory Platform will create new opportunities for social, economic, and political participation. Altogether (and with suitable rating and reputation mechanisms in place), this will promote better decisions and responsible actions. In particular, measuring the value of human and social capital and quantifying the 'social footprint' will help us to protect the social fabric on which our society is built, in a similar way as the measurement of the 'environmental footprint' has empowered people and institutions to better protect our environment.

In Conclusion

The FuturICT flagship project seeks to create an open and pluralistic, global but decentralized, democratically controlled information platform that will use online data together with novel theoretical models to achieve a paradigm shift in our understanding of today's strongly interdependent and complex world and make both our society and global ICT systems more flexible, adaptive, resilient, sustainable, and humane through a participatory approach. FuturICT is a big project: an unprecedented multi-disciplinary, international scientific endeavour requiring the collaborative effort of hundreds of scientists worldwide (see Box 3). The first practical results are expected 2–3 years after the project starts, with results and tools being made available to the public throughout the ten-year lifetime of the project.

This is an ambitious goal, but one that is within our reach. It will require advances in ICT, social science and complexity science. It will involve all levels and facets of society in developing and shaping the project's outcomes. Without an enterprise of this sort, the world is sure to fall increasingly behind the reach of orderly planning and management. Given the 21st century challenges that we need to tackle, it would be irresponsible *not* to undertake a project of this ambition. Everyone in the world is being affected by these new problems. They should all have a say in their solution.

Further Reading

Website of the Commission on the Measurement of Economic Performance and Social Progress, http://www.stiglitz-senfitoussi.fr/

Stiglitz and Sen's Manifesto on Measuring Economic Performance and Social Progress, http://www.worldchanging.com/archives/010627.html

D. Helbing and S. Balietti, How to create an innovation accelerator, *Eur. Phys. J. Special Topics* **195**, 101–136 (2011).

J. V. Henderson, A. Storeygard, D. N. Weil, *Measuring economic growth from outer space*, NBER Working Paper No. 15199 (2009)

P. S. Dodds, C. M. Danforth, Measuring the happiness of large-scale written expression: songs, blogs, and presidents. *Journal of Happiness Studies* **11**, 444–456 (2010).

S. Golder and M. W. Macy, Diurnal and seasonal mood vary with work, sleep and daylength across diverse cultures. *Science* **333**, 1878–1881 (2011).

Google Flu Trends. http://www.google.org/flutrends/

Planetary Skin Institute (http://www.planetaryskin.org/)

Digital Earth project *http://www.digitalearth-isde.org/*

D. Helbing, *FuturICT – New science and technology to manage our complex, strongly connected world*, preprint http://arxiv.org/abs/1108.6131

For further information, please consult the the following links:

http://www.futurict.eu: Project Webpage and English Media Response

http://www.facebook.com/FuturICT: FuturICT on Facebook

http://vimeo.com/futurict: FuturICT on vimeo

https://twitter.com/#!/FuturICT: FuturICT on Twitter

http://nextbigfuture.com/2011/09/vote-on-your-favorite-billion-euro-eu.html: FET Flagship Poll

FuturICT's Partners and Supporters

FuturICT is currently supported by about 1000 scientists worldwide (see www.futurict.eu). It involves Europe's academic powerhouses, such as ETH Zurich, University College London (UCL), Oxford University, the Fraunhofer Society, the Consiglio Nazionale delle Ricerche (CNR), the Centre National de la Recherche Scientifique (CNRS), Imperial College, and many more excellent academic institutions. More than five supercomputing centres support FuturICT. The project has letters of support from the OECD, the European Commision's Joint Research Center (JRC), research labs of Disney, IBM, Microsoft, SAP, Xerox, and Yahoo, international companies including banks, insurance and telecommunication companies, several regulatory authorities, and notable individuals such as George Soros. Furthermore, FuturICT has already managed to integrate many different research communities, and its leaders have a long track record of successful collaborations between scientists across disciplinary boundaries. FuturICT's supporters have also been involved in hundreds of successful projects with business partners.